Ecosystems

Since the first edition of this book was published in 1998, the role of ecosystems in understanding the environmental challenges faced by humankind has grown significantly. The ecosystem is *the* key concept in understanding the vital links between life and its environment that lie at the core of these challenges.

The second edition of *Ecosystems* explains the basic concepts that make up ecosystem theory and examines the ways in which the concept can help the investigation of environmental problems. The new edition has been revised and updated throughout to reflect the latest developments. It includes a new chapter on the world pattern of biomes and enhanced use of functional ecology in the assessment of ecosystem functioning. Ecosystem theory is first set in the context of functional ecology, itself a fundamental paradigm in contemporary ecology. Following a review of the historical development and refinement of the ecosystem concept, the authors explain how ecosystems function through analysis of the complex interactions between life and its physical environment. Using examples from around the world, the book addresses 'real world' problems. *Ecosystems* looks at the ways that this can be done at a range of scales, and analyses practical applications of the ecosystem concept. The increasing value of the ecosystem concept is demonstrated through its applications.

This updated edition explains the nature of the ecosystem concept, the functional roles of ecosystems, the ways in which it relates to functional ecology and its paramount value in the analysis of environmental problems. The book is illustrated throughout with boxes, figures, tables and plates.

Gordon Dickinson is Senior Lecturer in the Department of Geographical and Earth Sciences. **Kevin Murphy** is Senior Lecturer in the Division of Environmental and Evolutionary Biology, Faculty of Biomedical and Life Sciences, both at the University of Glasgow.

Routledge Introductions to Environment Series
Published and Forthcoming Titles

Titles under Series Editors:
Rita Gardner and A.M. Mannion

Environmental Science texts

Atmospheric Processes and Systems
Natural Environmental Change
Biodiversity and Conservation
Ecosystems
Environmental Biology
Using Statistics to Understand
    The Environment
Coastal Systems
Environmental Physics
Environmental Chemistry
Biodiversity and Conservation,
    Second Edition
Ecosystems, 2nd Edition

Titles under Series Editor:
David Pepper

Environment and Society texts

Environment and Philosophy
Environment and Social Theory
Energy, Society and Environment,
    Second edition
Environment and Tourism
Gender and Environment
Environment and Business
Environment and Politics, Second edition
Environment and Law
Environment and Society
Environmental Policy
Representing the Environment
Sustainable Development

Routledge Introductions to Environment Series

Ecosystems

Second edition

Gordon Dickinson and Kevin Murphy

Routledge
Taylor & Francis Group

LONDON AND NEW YORK

First published 1998
by Routledge
2 Park Square, Milton Park, Abingdon, Oxon, OX14 4RN

Simultaneously published in the USA and Canada
by Routledge
270 Madison Ave, New York, NY 10016

Second edition 2007

Routledge is an imprint of the Taylor & Francis Group, an informa business

© 1998, 2007 Gordon Dickinson and Kevin Murphy

Typeset in 10/12pt Times New Roman
by Graphicraft Limited, Hong Kong
Printed and bound in Great Britain
by TJ International Ltd, Padstow, Cornwall

British Library Cataloguing in Publication Data
A catalogue record for this book is available from the British Library

Library of Congress Cataloging in Publication Data
Dickinson, Gordon.
Ecosystems / Gordon Dickinson and Kevin Murphy. – 2nd ed.
    p. cm. – (Routledge introductions to environment series)
  Includes bibliographical references and index.
  Biotic communities.   I. Murphy, K.J. (Kevin J.)   II. Title.   III. Series.
  QH541.D535 2007
  577–dc22
                                                              2006016984

ISBN10: 0–415–33278–8 (hbk)
ISBN10: 0–415–33279–6 (pbk)
ISBN10: 0–203–40137–9 (ebk)

ISBN13: 978–0–415–33278–1 (hbk)
ISBN13: 978–0–415–33279–8 (pbk)
ISBN13: 978–0–203–40137–8 (ebk)

Contents

Series editors' preface
Environmental Science titles

The last few years have witnessed tremendous changes in the syllabi of environmentally related courses at Advanced Level and in tertiary education. Moreover, there have been major alterations in the way degree and diploma courses are organised in colleges and universities. Syllabus changes reflect the increasing interest in environmental issues, their significance in a political context and their increasing relevance in everyday life. Consequently, the 'environment' has become a focus not only in courses traditionally concerned with geography, environmental science and ecology but also in agriculture, economics, politics, law, sociology, chemistry, physics, biology and philosophy. Simultaneously, changes in course organisation have occurred in order to facilitate both generalisation and specialisation; increasing flexibility within and between institutions is encouraging diversification and especially the facilitation of teaching via modularisation. The latter involves the compartmentalisation of information which is presented in short, concentrated courses that, on the one hand are self-contained but which, on the other hand, are related to prerequisite parallel, and/or advanced modules.

These innovations in curricula and their organisation have caused teachers, academics and publishers to reappraise the style and content of published works. While many traditionally styled texts dealing with a well-defined discipline, e.g. physical geography or ecology, remain apposite there is a mounting demand for short, concise and specifically focused texts suitable for modular degree/diploma courses. In order to accommodate these needs Routledge has devised the Environment Series which comprises Environmental Science and Environmental Studies. The former broadly encompasses subject matter which pertains to the nature and operation of the environment and the latter concerns the human dimension as a dominant force within, and a recipient of, environmental processes and change. Although this distinction is made, it is purely arbitrary and is made for practical rather than theoretical purposes; it does not deny the holistic nature of the environment and its all-pervading significance. Indeed, every effort has been made by authors to refer to such interrelationships and to provide information to expedite further study.

This series is intended to fire the enthusiasm of students and their teachers/lecturers. Each text is well illustrated and numerous case studies are provided to underpin general theory. Further reading is also furnished to assist those who wish to reinforce and extend their studies. The authors, editors and publishers have made every effort to provide a series of exciting and innovative texts that will not only offer invaluable learning resources and supply a teaching manual but also act as a source of inspiration.

A.M. Mannion and Rita Gardner

Series International Advisory Board

Australasia: Dr P. Curson and Dr P. Mitchell, Macquarie University

North America: Professor L. Lewis, Clark University; Professor L. Rubinoff, Trent University

Europe: Professor P. Glasbergen, University of Utrecht; Professor van Dam-Mieras, Open University, The Netherlands

Note on the text

Bold is used in the text to denote words defined in the Glossary. It is also used to denote key terms.

Plates

Figures

Tables

Boxes

Authors' preface to the first edition

As is obvious from the title, this book is about ecosystems. A great deal has been written about ecosystems since the 1940s and there are some good academic textbooks about ecosystems. So, the reader is entitled to ask if we have anything new to say. We believe so.

The theme of the book is that ecosystems provide the best paradigm for the integration of the biotic and abiotic parts of the biosphere, and for the solution of real problems, as well as giving an adaptable theoretical base in the environmental and ecological sciences. It is written from the perspective that the ecosystem is *the* central concept in environmental science. We try to demonstrate this through a wide range of examples. Many of these include problems resulting from human impacts upon ecosystems. We think that the ecosystem concept can provide a very useful framework for the incorporation of the human dimension into biosphere functioning. We certainly do not imply that the population or community level analysis is of lesser value in ecology. But where *integration* and *large-scale* perspectives are needed, the ecosystem provides the best framework for research, whether this is purely scientific or directed towards resource management.

We begin by examining the development of the ecosystem concept. The concept has been much refined since it was initially proposed, incorporating advances in ecological and environmental sciences. Looking at the ecosystem in its current state of understanding, we first examine how the ecosystem functions. This functioning has two major subsystems: the flow of energy – an open system – and the cycling of materials – a closed system. This functioning is shown not only to be vital for the sustenance of life on Earth, but also to have a significant effect upon the abiotic parts of the biosphere. Thus the ecosystem gives us a means of describing the complex of reciprocal interactions between life and its physical environment. Though there are still problems relating to use of the ecosystem concept as a precise quantitative model, we contend that the way in which the ecosystem focuses on interactions can provide a useful framework for analysis of large-scale problems in the biosphere.

In our analysis of the ways in which the biological community subsystem functions, we use strategy theory (Grime 1979: see Chapter 2) as a means of explaining how organisms respond to both their biotic and abiotic environments. This theory, developed and widely applied since the 1960s, is still argued over by ecologists, but we think that it provides an excellent basis for developing models of the functional response of biota to the challenges posed by their environments. We illustrate this by examples taken from biomes from all parts of the world.

The book analyses both the biotic and abiotic subsystems which make up ecosystems. We do this to give a fuller understanding of the unifying position that the ecosystem concept occupies in environmental and ecological science. Too often, there is a lack of focus within environmental science. Research on environmental issues requires an integrating framework, which can give a coherence to the subject. In this book we show how ecology and environmental science can be linked via ecosystem studies.

We began working together in the early 1980s, when we began a research programme on the environmental changes which are taking place around Lake Nasser, the huge reservoir which has been created behind the Aswan High Dam, in southern Egypt. This work, which is continuing in the late 1990s, has looked at rapid change in natural ecosystems, modification of the whole physical environment and the creation of conditions which give a new resource base for human use. The ideas for this book and new research programmes came out of working together and long talks in the cool of desert evenings. We wish to acknowledge the considerable debt we owe to colleagues here and abroad. In particular the company and insights of Ian Pulford and John Briggs with whom we have worked in Britain, Egypt, Tanzania and Argentina are greatly valued. We have been fortunate in working in many different ecosystems. This has enriched our understanding of the world greatly, and not just the world of ecosystems.

The term ecosystem was first used by Sir Arthur Tansley. Though the concept has been developed, as the science of ecology has progressed, since his time, it retains the essence of what he proposed. That the concept has retained its level of utility in science is an indication of the underlying quality of the concept. It is worth noting too that Tansley, though educated in 'classical' botany, not only was a great pioneer in the new science of ecology, but also because of his interest in geography and geology may rightly be considered a pioneer of environmental science. The 'real' world, which is the subject of research in environmental science and ecology, has changed much since Tansley's time, and environmental problems are more serious, or at least are better defined than they were in the first half of the twentieth century. The use of the ecosystem concept as a means of understanding human misuse of the planet is a further measure of the importance and continuing academic strength and validity of the concept.

So what this book has to offer, which we think is distinctive, is the collaborative perspective of an ecologist and a physical geographer, based on more than a decade of working together. Our work has often been on 'real world' problems, and requiring practical as well as scientifically sound answers. We offer no prescription as to whether the ecological and environmental sciences are pure or applied. But many involved in these academic fields will work in applied areas, and it is impossible to strip out the role of human actions from ecosystems, throughout the biosphere. We have found the ecosystem concept to be a robust and adaptable one, for many purposes. As we have already said, it is not the only paradigm for the environmental sciences. However, when integration of a complex range of variables in the natural environment is involved, where human impacts, direct and indirect, are additional forcing factors, and where large spatial scales are involved, we contend that the ecosystem concept is an excellent way of approaching ecological issues.

Finally we offer our sincere thanks for the support of our families during this and our other collaborative ventures over the past decade. To our wives, Aileen and Fiona – both geography graduates, and now working as a computing analyst and town planner respectively – we gratefully acknowledge your forbearance and support, in this as well as our other projects. The opportunity to bounce ideas off you, and to have the sillier ones knocked down, has been a considerable asset to us. To our children, Rachel, Kathleen and Michael, we promise you a little more of our time in future. But we hope that you will have got something out of our absences. In researching together and writing this book, both of us have learned more about the world and some of its problems. If we can pass on some of this to other people, we hope that we can make a (very small) contribution to understanding these problems, and to keeping the world a good place for you and people like you.

Gordon Dickinson
Kevin Murphy
Glasgow, 1997

Authors' preface to the second edition

This second edition of *Ecosystems* has a number of changes from the first edition, which appeared nine years ago. We have tried to include some of the many important developments in ecology and environmental science that relate to the ecosystem concept. The functional ecology paradigm has a more prominent role in this new edition. We are both influenced by this approach to ecology, and believe that it is one of the most informative approaches in contemporary ecology. In response to helpful comments by reviewers, we have included a new chapter on biomes, and have extended and restructured the final chapters dealing with impacts on ecosystems. We hope that these changes make the book up to date and improve its utility for students. The ecological and environmental challenges that confront humankind grow with time. Environmental and ecological education is vital if we are to begin to solve the problems we all face.

In the period following the first edition, many colleagues have helped us a great deal. We must thank Mike Shand who produced many of the diagrams in this and the first edition, to the highest professional standards. In our research activities we would like to thank Nei, João and all our other friends in UEM and IAP, Maringá, Paraná State, Brazil. Our work with them on wetland and riparian rainforest ecology has been both a professional and personal career highlight. Our research students, especially Judith, Hazel and Gillian, have brought us new ideas and fresh insights. We have benefited from working with our colleagues in Glasgow in the Eurolakes programme, Colin, Jane and particularly Matt who did so much of the fundamental work. This project allowed us to carry out research with colleagues in Germany, France, Spain, Switzerland, Poland and Finland, which further enriched our experience. We would like to thank all colleagues in Glasgow and elsewhere who have helped us develop this second edition of *Ecosystems*. There have been too many to name individually, but they know who they are.

Finally, our wives Aileen and Fiona, and children Rachel, Kathy and Michael have continued to gives us vital support during long absences on fieldwork, and extended spells at the computer when we are back home. In dedicating this book to our families, we acknowledge the debt we owe to them for putting up with us. We have had all the fun while you have kept the show on the road.

Gordon Dickinson
Kevin Murphy
Glasgow, 2006

1 The nature of ecosystems

The biological world is one of great diversity and complexity. A systems approach is useful in helping us to understand the interactions between living organisms and their environment (which includes the biotic environment of other living creatures). The concept of the ecosystem provides a way in which the functioning of the biological world and its interactions with the physical environment can be understood. The ecosystem concept is useful in resource management and as a basis for predictive modelling. This chapter covers:

- **Complexity of the biological world and its physical environment**
- **Development of the ecosystem concept**
- **System theory, ecology and ecosystems**
- **Abiotic and biotic environment of ecosystems**

How this book approaches the complexity of the biological world and its environment

How can we make sense of the complex and constantly changing interactions between the living world, with its myriad species and individuals, and the multifaceted and dynamic environment which life inhabits? In this book we examine this basic question, starting from the idea of the ecosystem as the basic unit of living organisms in the environment. Understanding how ecosystems operate, and how they support the existence of groups of organisms, is not just a question of scientific interest. At a gathering pace since the 1940s, there has been increasing concern about harmful effects caused by human actions on the planet's life support system. Although concerns were, at first, confined to a small group of scientists and environmental activists, it is now a global issue at the top of the international political agenda. Exactly what has occurred and what may happen in the future is not clear. However, most informed people agree that at best the consequences may be uncomfortable for humankind, and at worst may be catastrophic.

The ecosystem concept is fundamental to examination of human impacts on life on Earth. It provides a way of looking at the functional interactions between life and environment which helps us to understand the behaviour of ecological systems, and predict their response to human or natural environmental changes.

In this chapter we describe the evolution of the ecosystem concept, and its contemporary definitions. Many people have some idea of what is meant by the term *ecosystem* (see Definition Box).

Definition

Two definitions of the term *ecosystem*

- 'An energy-driven complex of a community of organisms and its controlling environment' (Billings 1978).
- 'An ecosystem is a community of living organisms together with the physical processes that occur within an environment' (Pullin 2002).

These two definitions, nearly 25 years apart, provide consistent statements on the key attributes of ecosystems. These key attributes are directly related to the concepts of functional ecology which are used in this book. In particular, interactions between the physical environment and organisms, and between organisms and other organisms direct the evolutionary trends of competition, tolerance of stress, and tolerance of disturbance. These interactions are central to the functional processes specified in the definitions of ecosystems.

Ecosystems can be analysed using the concepts of system theory. This approach provides definitions and general rules which allow very complex structures to be understood and predicted. When allied to mathematical modelling techniques, system theory provides the framework for a highly effective general approach to the study of ecosystems. We examine below some of the main issues in system theory, and relate these ideas to the ecosystem concept.

Ecosystems are found throughout the **biosphere** (Flanagan 1970). The biosphere is the zone in which life is located, in a shell around the planet. If abiotic environmental life support systems are included, this zone is sometimes referred to as the **ecosphere**. Within the ecosphere, ecosystems exist at spatial scales from a crack in a rock (see Chapter 5 for more on the endolithic ecosystems of Antarctica) to rainforest or oceanic ecosystems, covering areas of thousands of square kilometres (see Chapters 3 and 7). Sometimes the boundaries of ecosystems coincide with natural spatial features, such as an island or a type of vegetation, such as a forest. However, ecosystem boundaries may be defined by purely human criteria, such as a national or state boundary. Ecosystems may even be artificially constructed in the laboratory. Biomes are the largest-scale units which depict the global pattern of the distribution of vegetation in the biosphere. This pattern is generally related to current and recent climatic conditions, and contrasts with the pattern of zoogeographical realms, which relate to barriers to dispersal and to the outcomes of continental drift. As the most important element in any ecosystem is its vegetation, which provides the input of energy into the whole system, we examine the global patterns of biomes in more detail in Chapter 7, and relate these patterns to elements of functional ecology.

The biosphere extends from at least 0.5 km below the floor of the ocean into the atmosphere. Life has been detected up to 6.5 km above the Earth's surface. This is close to the **tropopause**. Thus the biosphere is no more than 20 km thick, 0.3 per cent of the planetary radius. However, as far as we know, it is the home of all life (though see our speculation on the possibilities of life elsewhere in the Solar System in Chapters 3 and 5).

Ecosystem functioning is the main theme of this book. In Chapters 2, 3 and 4 we outline the functional interactions between energy and materials in ecosystems, and the way in which these support life in ecosystems. Understanding the operational and support functions of ecosystems (how they work and what they do) is vital to the use of the ecosystem concept for predictive purposes (for example, understanding the potential impacts of global warming: see Chapter 11). The energy and material subsystems are analysed individually in Chapters 3 and 4. In reality these are intimately interrelated in the operation of ecosystems. Most of the materials which are required to construct living organisms are in relatively short supply within the boundaries of the biosphere. Cycling of these materials by ecosystems is thus a critical part of the whole life support system of the planet.

Ecosystems interact in a variety of ways through their **biotic** and **abiotic** components. Chapter 5 analyses the general response of ecosystems to stresses imposed by different physical environments and human activities. Seasonal and other temporal changes in ecosystem characteristics are an important variable influencing the intensity and timing of environmental stress affecting ecosystems. Natural change is a normal feature of the functioning of the Earth's environment. Sometimes the disturbance produced by such change can be massive in its effects, resulting in conditions unfavourable to all or most members of the pre-existing biological community. Extreme examples include the effects of a major meteorite strike (such as the 'dinosaur killer' thought to have been responsible for the mass Triassic extinction suggested by Gould as far back as 1980) or a major volcanic eruption (see Chapter 2 for an example). Much more common are the effects of disturbance caused by grazing organisms for producer species like plants. Some of the most important aspects of ecosystem response to disturbance are discussed in Chapter 6. But much of the functioning of ecosystems is shaped by response to interactions between the various biological populations which make up the community structure of ecosystems.

Functioning ecosystems always change through time. The dynamic nature of ecosystems operates over time scales ranging from daily to geological time. One of the most important dimensions of this interaction is competition between individual, and populations of, organisms. This is analysed in Chapter 7.

Change to ecosystems may be caused by human actions. One of the issues that give rise to the greatest concern among scientists concerned with the environment, and among the public at large, is the effects that humans are having upon ecosystems and their functioning. These human impacts act at various scales and with varying severity. Analysis of selected examples in Chapters 9 to 11 critically assess what effects human impacts may have, and how serious these threats are to ecosystem function. One of the most difficult problems facing environmental science is diagnosing the nature of environmental change. Not only is the extent and rate of change often hard to detect, and even harder to predict, but it may also be very difficult to distinguish between those components of change which are a part of natural environmental and ecosystem dynamics, and those which are a result of human impacts. Yet unravelling all of these issues is vital if ecosystem function is to be sustained, and irreparable damage to the biosphere avoided. These problems are discussed more fully in Chapters 9 to 11.

The ecosystem concept and the biological world

The ecosystem concept provides a convenient means of structuring and understanding the highly complex system which is our world. Even now a significant proportion of living organisms on this planet remains undiscovered and unclassified. It is likely that there are whole ecosystems which as yet remain unknown (especially in the oceans).

If the different kinds of organisms present a formidable array of forms and functions, this complexity is added to by the fact that, to a greater or lesser extent, each individual organism is different from all others of the same kind. Some living organisms do not conform to this rule by reproducing asexually, but individual distinctiveness is one of the keys to survival. An essential element of life is that species must exist in numbers sufficient in both time and space to be able to support breeding at a level which will

Box 1.1

Hierarchy of life: level of integration and links

Hierarchy	Level of integration		Links
Biosphere	↑ ↑	↓ ↓	Macro-scale environment
Biomes	↑ ↑	↓ ↓	Meso-scale environment
Ecosystems	Increasing complexity of organisations ↑	↓ ↓ ↓ ↓	Defined envelope of environment and biota conditions
Functional groups	↑ ↑ ↑ ↑ ↑	Decreasing number of individual organisms ↓	Sets of environmental pressures within tolerance range of species making up functional group
Communities	↑ ↑ ↑ ↑ ↑ ↑	↓ ↓ ↓ ↓ ↓ ↓	Sets of environmental pressures within tolerance range of species making up community
Populations	↑ ↑ ↑	↓ ↓ ↓	Other populations and micro-scale environments
Organisms	↑ ↑ ↑ ↑	↓ ↓ ↓ ↓	Other individuals, of the same and other species, and micro-scale environments

replace individuals lost through death. These groups of individuals are called **populations**. Populations form the next step in the hierarchy of life after individuals. Groups of populations which occur together in defined locations form recognisable **communities** of species. Where these communities are adapted to similar combinations of types and intensities of environmental pressures (in one or more geographically distinct locations on the planet's surface) they form **functional groups** of species. One or more functional groups of organisms (sometimes many), together with a defined set of abiotic environmental conditions, form an ecosystem. Groups of ecosystems which share broad environmental characteristics are termed **biomes**. Finally, the whole global assemblage of biomes comprises the biosphere. The hierarchy is shown in Box 1.1. The distribution of land biomes is shown in Figure 1.1. Chapter 8 considers the relationships between biomes and their abiotic environment in detail.

To understand ecosystem functioning we must appreciate what each level of organisation involves, how it relates to levels above and below, and how the whole structure is integrated. At the level of the individual, an organism will grow and may reproduce. Its

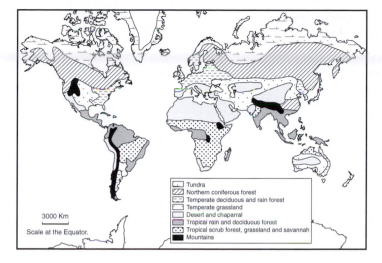

Figure 1.1 Distribution of land biomes

genetic characteristics can be transmitted from generation to generation, and through the process of natural selection will help to ensure the survival of the species. Over numerous generations this process may result in the evolution of a new species which has a specific ecological **niche**: its functional role with respect to its biotic and abiotic environment. The individual interacts directly with other individuals of the same and other species, through competition and predation. Any individual organism is also profoundly affected by its controlling abiotic environment. The population, comprising a number of individuals of the same species, contains a wider range of genetic information than any individual. The community is the aggregate of all biological populations in a defined area. Plant, microbial and animal communities are usually distinguished. Populations respond to the environment by adaptation, and all individuals within the population are in competition for resources to sustain life. Populations interact with other populations within communities to form functional groups, in response to biotic interactive pressures, such as consumption and competition for biological resources like water and light, and also to abiotic stress and disturbance pressures on survival and reproductive success.

One type of relationship which is of importance to the understanding of ecosystems is the **trophic structure** of the community (Figure 1.2). Trophic structure may be defined as the structure of energy transfer and loss between different populations in the community. Every population belongs to a particular trophic level. This is a statement of its position in the energy transfer structure of a particular community. This is important in understanding ecosystem function, and trophic structure is characteristic in many general types of ecosystems, such as lakes or deciduous forests (Odum 1971). Trophic levels and trophic structure are explained more fully in Chapter 3.

Environment of the biological world

The abiotic environment, often termed the physical environment, consists of a series of complex, interactive energy-driven systems. Those with which we are concerned function in the biosphere. This term was first used by the Russian mineralogist V.I. Vernadsky (1863–1945) as a means of providing a holistic view of nature, including the abiotic environment. It is by no means coincidental that this concept first emerged in

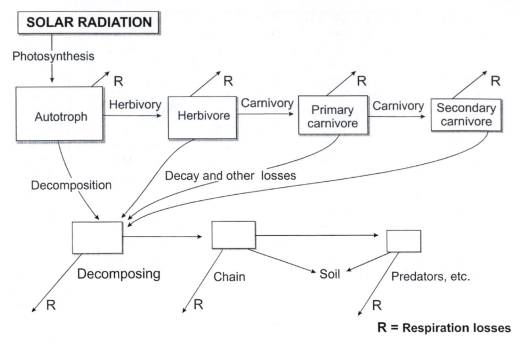

Figure 1.2 Trophic structure and energy flow in an ecosystem

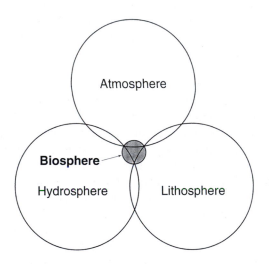

Figure 1.3 Physical environment of the biosphere

Russia, immediately following the Bolshevik Revolution, when perspectives integrating life, including human activities, with the physical environment, were fashionable (Bowler 1992). The systems of the physical environment are influenced, and in some cases controlled by events and factors which lie beyond the biosphere, but these issues are beyond the scope of this book. Readers requiring further information on physical environmental processes are referred to other titles in this series.

Those parts of the abiotic environment which act on the biosphere are shown in Figure 1.3. The biosphere, with all its component ecosystems, is located at the junction of three terrestrial 'spheres' or shells around the planet: the **atmosphere**, **hydrosphere** and **lithosphere**. Like the biosphere, these shells are highly dynamic, and change in the physical environment is normal. The dynamic properties of the physical environment are driven by energy, and most of this energy is solar radiation. In the case of large-scale processes, affecting the Earth's crust and operating at geological time scales, energy is derived from the vast amount of heat which the Earth's core still contains. Tidal energy is derived from gravitational interaction between the Earth, moon and sun. However, the majority of environmental processes, such as weather systems, the hydrological cycle, ocean currents or surface erosion, are almost exclusively driven by solar radiation.

The dynamic nature of the physical environment is not the only reason why ecosystems are dynamic. Organisms must react to the challenges and opportunities of the physical environment as well as interacting with other organisms. Ecologists use the terms **habitat** and niche to describe how organisms relate to their environment. In a particularly good metaphor, habitat has been described as an organism's 'address' and niche as its 'profession' (Odum 1993). In other words an organism's habitat is the geographical location at which that organism lives, including the physical environmental characteristics of that location. Depending on the level in the biological hierarchy which is under study, habitat may refer to a very limited area, measured in a few square metres, for an individual organism, to subcontinental regions extending over thousands of square kilometres, for communities. Variations in habitat scale lead to the term *micro-habitat* being used for locations and environments influencing a single or small group of individuals. Niche is rather more complex and a number of types of niches have been defined. The notion of niche, first used by such pioneers of ecology as Charles Elton (1927), described the relationship between habitats and behaviour or response of species, particularly in respect of competition and predation or consumption. These relationships concerned species' functional interrelationships, in which the physical environment was a kind of stage for biological activity. Gause (1934) and Lack (1947, 1954) used niche relationship concepts to investigate competition and evolutionary diversification of species (Ricklefs 1990).

A more specific definition used as studies of ecological energetics developed in the 1940s was trophic niche. This is the relationship between an organism or population and other members of its community in terms of energy flows (Odum 1971). This allowed a more precise statement to be made. Although this definition permitted quantitative data to be used in describing niche, its scope was limited. It did not include a direct statement on the nature of physical environment–species interaction. This is vital, though it leaves out such factors as the impact of seasonal patterns of climate upon plant productivity or the differing growth responses of different plant species to variations in soil conditions.

A major step forward was made by Hutchinson (1957). He envisaged the environment as being a series of dimensions, along which the niche of any species could be located. This is easy to visualise with only two or three dimensions, which can be represented as an axis in real space. Hutchinson stated that there should be as many dimensions as there were measurable ecological factors. This cannot be represented in real or Euclidean space, but can be constructed abstractly by mathematics. Hutchinson's view of the niche was termed hyper-volume or *n*-dimensional niche. This definition allows precise definition of the relationship between any organism and its total environment.

Use of Hutchinson's concept advanced ecological science through the promotion of research into two problems which follow from the definition. First, as there are many ecological dimensions involved in the ecology of any species, how can the most important environmental factors or dimensions be identified and assessed? Second, how can the way in which a species occupies a space or range along an environmental dimension be assessed? In most cases there will be a range of values for any environmental factor for a particular species, within which the species may be found. Generally the optimum conditions for that species will be towards the middle part of the range, so that more individuals will be close to this central point, while some individuals will be found towards the limits of the range (see Chapter 2). The relationships between abundance and ecological dimensions are of importance in understanding its overall ecology. However, as the physical environment is dynamic, individuals and species must be able to tolerate a range of ecological conditions along each niche axis. As environment changes constantly in time, according to patterns of variation in the physical environment, such as seasonal climatic conditions, wide or narrow tolerance of these variations is an important aspect of a species' ecology. Both of these problems have been attacked with vigour in the research of ecologists since the 1960s, and understanding of community ecology and ecosystem function has advanced greatly. This work has been based on the adoption of mathematical techniques in ecological science, and greatly advanced by the development of powerful statistical techniques and appropriate computer power to carry out the work.

Development of the ecosystem concept

Early ideas

Modern ecological science and the study of ecosystems grew from early interest in what was called natural history. Gilbert White's *The Natural History of Selborne* (1789), a classic study of plant and animal life in the area around an English village in the eighteenth century, is an early example of this work. Much early study of the living world was spurred by practical concerns such as agriculture and sylviculture. Exploration, which proceeded at an accelerating rate through the nineteenth century, often included a scientific dimension through collection of specimens of plant and animal life. The famous voyage of *HMS Beagle*, which took place between 1831 and 1836, and during which Charles Darwin made the observations which led to his evolutionary theory, is one of the best examples of this. Knowledge of the living world was systematised by classification of new species as they were discovered, and some basic information about habitats was often recorded too. Natural history became a popular hobby for the growing numbers in the educated middle classes, and the endeavours of countless dedicated amateurs as well as a few pioneer professionals advanced the quantity of knowledge about the living world considerably. However, it was not until the end of the nineteenth century that basic ecological questions were asked. Two general themes were identified. First, stemming from the descriptive classification of individual species came the notion that plants and animals lived together in distinctive and recognisable assemblages, or what are now termed communities. These assemblages were found in particular locations or habitats, and influenced the patterns of distribution of species. Second, and following from this, there were interrelationships between communities, particularly relating to the ways in which plant growth, competition, consumption and predation affected the types and numbers of species found in the community.

At the beginning of the twentieth century, ecology as a recognisable academic discipline began to appear. The historical development of ecology and environmental sciences is analysed by Bowler (1992). The two issues of identification of assemblages of plant species, and of the interrelationships between them and their environment, were the foci for research. In the United States, H.C. Cowles and F.E. Clements investigated the development of vegetation, through a series of stages, within which community assemblages were similar (Cowles 1899; Clements 1916, 1936). Both worked on sand-dune vegetation, in which patterns of plant communities are often very distinctive, and vegetation and environmental changes occur over short spatial distance, following major ecological gradients. Clements developed a general theory of vegetation **succession**. This was based on the notion that as a community developed, it modified its physical environment in such a way as to produce a new set of environmental conditions which were less favourable to the initial community, which was then replaced by a new community. The stages in succession were termed **seres**, and a final stable condition was eventually reached. This he called **climax** vegetation. Clements believed that the nature of climax vegetation was determined by climatic conditions alone, and that other ecological factors were of secondary importance. Successions developed from bare new land surfaces, such as, in the case of sand-dunes, the upper beach above the normal, daily tidal range, or a land surface emergent from beneath a retreating glacier. Succession would also occur following the removal of a pre-existing vegetation cover by such agencies as fire or erosion. These sequences were termed secondary successions, and such patterns were frequently associated with human actions. Clements saw the development of vegetation towards the stable end of the climax as similar in development to that of the growth of an individual organism, and likened the community to a 'super-organism'. This theory, though influential and widely accepted, was challenged by other workers. Although evident in the case of sand-dunes, evidence from other types of vegetation, such as temperate forests, led some researchers to the view that vegetation did not follow a sequence of development, and that generally recognisable, related climax communities did not exist.

This alternative view was that vegetation was composed of unique combinations of numbers of individuals of different species. Each tract of vegetation was functionally unrelated to all others, except that individual species happened to grow in a particular location because of adaptation to that environment. The principal advocate of this perspective on vegetation was Gleason (1926), who argued that while plant communities, which he termed associations, could be convenient abstractions, they had no functional reality other than the interaction of individual species and consumption by herbivores. This was a clear rejection of the 'super-organism' concept. The controversy about the nature of vegetation was to continue for several decades, and particularly to focus on the issues raised by Clements and Gleason. As more powerful analytical techniques became available, increasingly sophisticated investigations of developmental processes were made. Clements' view that the climax is exclusively determined by climatic conditions, the monoclimax, has been modified to a polyclimactic perspective, in which one or more other environmental factors may influence succession. Furthermore, it has been shown that, in detail, communities at any stage of seral development show internal variations, which relate to **stochastic** processes, or patch dynamics, controlled by local environmental and competition factors. The contemporary perspective on succession is that there is a wide range of processes which control the development of succession. Connell and Slayter (1977) proposed two theories of causation of succession. The first was the so-called 'facilitation model'. This is similar to Clements' original ideas about succession, in that it envisages that the primary cause of seral development is change in physical conditions produced by plants at an earlier seral stage. The second theory, which was favoured by the authors, at least in the case of secondary succession (i.e. succession which starts from a surface from which vegetation has been wholly or partially removed), was termed the 'inhibition model'. In this, species resist invasion until they are replaced by competition, predation and disturbance.

E.P. Odum (1983) suggested that in the course of autogenic succession, not only are there increases in the rate and efficiency of nutrient cycling and energy flow, but also there are trends to increases in symbiosis and ecosystem resistance, and a decrease in

---

Box 1.2

Gaia hypothesis

The Gaia hypothesis was developed by James Lovelock in 1979. Having made a high scientific reputation, and achieved financial independence through his development of the electron capture detector, a key device in environmental analysis, he turned his attention to a unified view of earth and life sciences. He put forward the idea that all the environmental and ecological systems of the earth were linked in a complex but self-regulating system which evolved over geological time periods. He proposed that the atmosphere of the earth had been changed by life; this produced a climate which was favourable to life. There is evidence to support this notion. If photosynthesis did not exist, there would be much more CO_2 in the atmosphere and the surface temperature of the earth would be much hotter than it presently is. Lovelock argues that it is life which has shaped the atmosphere and its climatic properties, and that life acts as a stabilising, negative feedback control on the climate. He illustrates this evocatively with his model, 'Daisyworld', in which the numbers of dark and light coloured daisies regulate a planet's temperature (Lovelock 1988).

The Gaia hypothesis has been controversial from its appearance. When Lovelock proposed it he thought that he would be criticised from church pulpits. Instead it is members of the scientific community who have been the most severe critics. Neo-Darwinists maintain that Gaia is an organismic theory, which accords with neither evolutionary theory nor the evidence of evolutionary trends. However, some scientists think that at least the Gaia theory has helped to illuminate the nature of interactions between life and its environment. People who are interested in ecological science and ecosystems should read Lovelock, and make up their own minds.

---

ecosystem resilience. Such ideas are controversial. Many biologists (not just the vocal group of contemporary neo-Darwinists) vigorously reject any theory which appears to have organismic underpinnings. Odum's views on the nature of succession are seen by his critics as being close to this, and a contradiction to the established primacy of the 'trial-and-error' control of natural selection. Nevertheless, the concept of succession remains important in understanding interactions between organisms and their environment. These issues are discussed in Chapter 6. This issue is also explored more fully in the discussion of the Gaia hypothesis in Box 1.2.

The second fundamental ecological problem which was receiving attention during the early part of the twentieth century was that of the nature of functional relationships within biological communities. During the 1920s the English biologist Charles Elton conducted field research in the tundra of the Norwegian island of Spitzbergen. This area, which is located almost 80°N, is subject to a severe climate producing intense climatic stress on plants and animals living there. The issues raised here are discussed in more detail in Chapter 5. The island has a simple biological community structure. A year-round complete ice cover is prevented only by the moderating effects of the Arctic Ocean around the shores of the island, where limited and specialised vegetation cover develops. The simplicity of the community structure and the degree of control exerted by its harsh climatic environment made this a suitable area for Elton's pioneering studies. Elton's work focused on analysing the patterns of consumption between the plant and animal populations of the tundra. This was the basis of his subsequent theoretical proposal of the concept of the **food chain** (Elton 1927).

This simple idea was based on Elton's view that, as the survival of animals is based on food consumption, the feeding patterns of each population were among the most important aspects of biological community structure. He pointed out that plants, or more properly **autotrophs**, played the fundamental role in any food chain since only autotrophs could synthesise organic materials ('food') from inorganic inputs, utilising solar radiation. The function of all populations in the community could be identified by their feeding interrelationship. This was termed the trophic structure of the community. As is discussed further in Chapters 2 and 3, all autotrophs are at the first trophic level. Primary consumers, or grazers, are at the second trophic level, primary carnivores at the third trophic level, secondary carnivores at the fourth trophic level, and so on. The direct above-ground food chain in terrestrial ecosystems is paralleled by a sub-surface soil or detrital consumption food chain. The substance of Elton's theory led to interest in how energy was transformed and transferred through biological communities, or **ecological energetics**. The simple food chain concept has been replaced by the notion of a **food web**, in which consumers may obtain food from populations at different trophic levels (see Chapter 3).

Energetics studies have enabled ecologists to gain a better understanding of the ways in which populations respond to external environmental stresses. An example of this is response to seasonal variations in energy flow in communities. Seasonality can be defined bioclimatically as the occurrence of an unfavourable season for plant growth, due to low temperatures or water deficit. Seasonal patterns of variation in primary production by plants are related to climatic controls and to variations in the numbers of consumers, related to mortality and migration. By the 1930s the notion that the community comprised an interactive group of species was becoming a significant element in mainstream ecology. When taken together with the advances in research into community composition and dynamics, this led to a major advance in conceptualisation of the ways in which organisms and their environment interacted.

Although there was still support for the kind of 'super-organism' view of biological communities which Clements had initiated, it was a reaction to this notion that saw the first use of the term **ecosystem** by the English ecologist Sir Arthur Tansley. It should be noted however that as early as 1877 the German scholar K. Möbius proposed a rather similar notion which he termed 'biocoesis', a term still in use in some non-English literature today. Tansley was not only a major figure in the development of plant ecology but also a great populariser of the subject. His beautifully written book *The British Islands and their Vegetation* (Tansley 1949b) gave an authoritative and evocative account of British vegetation. He was also a great character. Box 1.3 tells a little more of the life of this great scientist. Tansley's ecological work began with experimental verification of what had long been suspected about competition between plant populations. He showed that though species could tolerate unfavourable environmental conditions when grown in isolation (in his experiment soil reaction), when grown together, the species best adapted to the specific environmental factor under investigation would oust the species less well suited to that environmental condition. This led Tansley to the view that the 'super-organism' notion was not valid, but that the community and its environment existed in a 'system in the sense of physics' (Tansley 1935). In this system, a complex of interactions between organisms and their environments defined community structure and function. This he termed 'the ecosystem'. The ecosystem included both communities of organisms and their physical environment, and organisms interacted with this abiotic

Box 1.3

Sir Arthur Tansley: a founder of modern ecology

Arthur Tansley (1871–1955) is one of the main figures in the development of modern ecological science. He first brought into use the concept of the ecosystem, and he undertook critical research into the niche concept. He was founder and first president of the British Ecological Society, and founder and first editor of two of the most important scientific journals devoted to ecology, the *New Phytologist* and the *Journal of Ecology*. Tansley grew up in a comfortable middle-class home. Supported by his parents, he developed a great interest in science, at a time when most young men of his background studied the humanities or entered the professions.

Tansley studied at University College, London, and Trinity College, Cambridge. His early university career was spent in University College, Cambridge. He was appointed to the chair of Botany in Oxford University in 1927, which he held until he retired ten years later. He was elected FRS (Fellow of the Royal Society) in 1915 and knighted in 1950. His distinguished academic career was accompanied by a life-long interest in adult education through the Working Men's College. He travelled widely, conducting fieldwork in many different environments.

He corresponded with many of the other seminal figures in the embryonic discipline of ecology, including F.E. Clements, and H.C. Cowles, with whom he had a long friendship. He was greatly interested in the work of Sigmund Freud, the psychologist, and studied with him in Vienna in 1923. He was interested in the academic disciplines of geography and geology. Besides a considerable output of scientific literature, he wrote for a wider audience, with great skill. *Britain's Green Mantle* (published in 1949) is a good example of the way in which he could draw environmental and human factors into the analysis of vegetation. He was a highly regarded teacher, influencing the whole generation of ecologists who followed him.

But he was a very human person. He liked entertaining, food and wine. By no means the only ecologist with these foibles, he enjoyed fast cars, though his students wished he did not. He is now remembered as a founder of modern ecology, and the father of the ecosystem concept. We should remember that he had much wider interests and was a man of great personal qualities too. Scientists are people and understanding what sort of people great scientists were adds to the appreciation of their work.

For more about Tansley, read the affectionate tribute to him by his pupil, Sir Harry Godwin (1977) in the *Journal of Ecology*.

environment, as well as the biotic environment produced by the other populations in the ecosystem. It is interesting to note Tansley's words 'system in the sense of physics', for at this time the first ideas about systems as structures which were found widely in the real world were being developed. This body of theory showed that these complex natural and human-constructed systems could be analysed through a novel application of mathematical and logical theory. This was termed 'system theory', and, as is discussed in the next section of this chapter, is highly relevant to the ecosystem concept and its development up until the present. The final section of Chapter 6 reviews the development of the use of the ecosystem concept in the context of the functional ecology of vegetation dynamics and spatial patterns.

A review of ideas about the ecosystem concept

The use and definition of the term 'ecosystem' by Tansley was followed by substantial progress in understanding how ecosystems function. Initially this was based on research into ecological energetics. Although the first studies into ecological energetics by Lotka in the 1920s pre-dated Tansley's theories, and gave a thermodynamic structure to the ecosystem which fitted the developing ecosystem concept (Lotka 1925), little attention was paid to his work at the time. Lotka developed a simple energy cycle system in which input of solar energy was balanced by heat output, following the cycling of energy as foodstuffs through the various trophic levels of a simple ecosystem.

It was not until the work of Lindeman (1942) nearly two decades later that energetics became a major area in ecological research. Lindeman defined the term **trophic level**, and pointed out that decreasing amounts of energy were available at successive trophic levels due to heat losses at each trophic level. These heat losses, which balanced the input of solar radiation in conformity with the laws of thermodynamics, resulted from organisms' use of energy in metabolic processes, such as respiration. The laws of thermodynamics state that energy cannot be created or destroyed, and that therefore in a system there must be a balance between input and output of energy. Thermodynamics also state that the ultimate fate of energy is to be transformed into heat, the energy condition with the highest entropy state. **Entropy** may be thought of as the degree of disorder in the total energy content of a piece of matter. Biological materials carry energy in a condition of relatively low entropy in chemical bonds in compounds. This energy, which is consumed in food, is broken down by organisms' metabolism to accomplish life functions (e.g. growth, reproduction) and then is lost to the atmosphere as heat, and ultimately to space as part of the out-radiation from the Earth. Lindeman's work showed how a major part of ecosystem function could be measured and modelled.

By the mid-twentieth century there was a clear idea of structure and energy flow in ecosystems. E.P. Odum, probably the most influential ecologist working at the ecosystem level since the 1950s, took energy cycling further by demonstrating that the energy cycle was paralleled by a nutrient cycle (Odum 1953). Over a hundred years before, the German chemist Liebig had shown that plant growth was controlled by the nutrient element which was in shortest relative supply. Figure 1.4 shows that plants have a

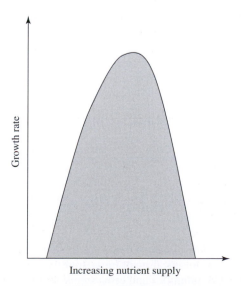

Increasing nutrient supply

Figure 1.4 Relationship between nutrient supply and plant growth rate minimum requirement, an optimum intake and a maximum tolerance for any nutrient (Liebig 1840). **Nutrients** are the chemical elements which are required to build organic matter. All green plants require specific amounts of each nutrient. Too little or too much will inhibit or even prevent plant growth. Odum showed that as nutrients in the **available** form – that is, in a state and location in which they may be used by autotrophic plants – are in limited supply throughout most parts of the biosphere, cycling of these nutrients is vital to sustain energy flow in ecosystems, and thus life on Earth. Nutrients and nutrient cycling are examined more fully in Chapter 4.

Since Odum's influential study, much ecological research has focused on the ecosystem. Better methods of measurement in the laboratory and the field, better application to ecosystem analysis of theories in physical and biological science, more effective use of mathematical and statistical techniques, allied to the exponential growth in computational power of this period have all contributed to a better understanding of the ecosystem. The system approach has been extended by use of the philosophy of general system theory and the methods of system analysis by numerous ecologists (e.g. Jeffers 1978; Odum 1983). The method was employed widely during the research programmes of the International Biological Programme (IBP) of the 1960s and 1970s, with, however, mixed success.

This has led some ecologists to question the value of the ecosystem concept, particularly as a primary research tool. Some antagonists maintain that the population is the best level for primary research, and that at best the ecosystem is a useful illustrative concept. A further problem has been identified by Odum. He has investigated the notion that ecosystems do indeed have organismic properties. This has been a controversial notion from times of the earliest ecological research. Odum and others (e.g. Margalef 1968), have looked at development, stability through regulatory feedback processes or homoeostatic mechanisms. This may imply that it is self-regulating in the way that an organism regulates its own internal environment, and may even grow old in the way an organism ages. These are highly controversial ideas, very difficult to test, and rejected by many ecologists.

A further area which has advanced thinking about ecosystems is the growth of scientific and popular concern about environmental and ecological degradation. The unifying and integrative nature of the ecosystem concept has seen its application to problems both practical and theoretical. Ecosystem theories have been applied widely in the development of conservation management strategies (e.g. Usher 1973). The important ecosystem–climate links have been incorporated into research into global climatic change (e.g. Schneider 1994). Again there have been critics of these approaches, especially in basic research work. However, few ecologists find fault with the way in which the concept has served to advance knowledge of ecosystems in the academic bases of tertiary-level education (e.g. Odum 1993) and in the popular media, which has reinforced the general public's concern for and knowledge of the richness, diversity and vulnerability of life and the environment of our planet (e.g. Attenborough 1979).

This book uses functional ecology as a key element in understanding ecosystems and their functioning. The perspective on functional ecology used here is based on the premise that plant strategies exist, and as a consequence functional groups may be identified. Plant strategies have been succinctly described as 'groupings of similar or analogous genetic characteristics which occur widely among species or populations and cause them to exhibit similarities in ecology' (Grime 2001, xxvii). **Plant strategy theory**, sometimes called CSR theory (see Chapter 2), was originally purely conceptual, but is now based on a large body of empirical evidence. This has involved measurement of anatomical and physiological plant traits, multivariate analytical techniques and testing of predictions. Communities and ecosystems are the focus of this empirical work. A recent study which analyses the contention that plant traits drive ecosystems, and which is based on a range of studies from three continents, provides compelling evidence for the application of strategy theory to the study of ecosystem functioning (Diaz *et al.* 2004). Recently, the analysis of relationships between **biodiversity** and ecosystem functioning has been a major theme in ecological research (Loreau *et al.* 2002). This has been concerned with the effects of changes in biodiversity, such as extinction of a particular species at a specific location or all locations within the biosphere on ecosystem trajectory. Ecosystem trajectory means predictable change in ecosystems characteristics which are controlled by its biotic components and their interaction with the abiotic environment.

System theory, ecology and ecosystems

At the beginning of this chapter we asked how it might be possible to make sense of the complexity of interactions between the living world and the environment. The discussion of the evolution of the ecosystem concept in the two previous sections points in the direction of the development of increasingly rigorous and mathematical analysis of the interactions between the living world and its environment. To a considerable extent this is based on system theory and systems analysis and modelling. System theory, sometimes termed 'general system theory', and systems analysis are sometimes thought of as being one and the same. This is incorrect. Properly, system theory is a body of theory in the realms of philosophical logic and of mathematics which concerns the nature and properties of those structures and are defined as systems. All terms in this section printed in bold are included in Box 1.4, which gives definitions of key system concepts. Systems analysis is the development of techniques of analysis of systems and the application of these techniques to building **models**, or mathematical representations of systems. The development of ideas about systems, which may be termed 'systems science', and which includes both theoretical and practical perspectives, was related initially to advances in physical sciences and engineering, but since the 1950s systems science has been applied to a very wide range of problems and disciplines, including business and the humanities. As previously noted with respect to the IBP (International Biological Programme of the 1960s and 1970s), the systems approach has been a significant element in ecological

---

Box 1.4

System theory definitions

system

Any collection, grouping, arrangement or set of elements, objects or entities that may be material or immaterial, tangible or intangible, real or abstract to which a measurable relationship of cause and effect exists or can be rationally assigned.*

system boundary

A physical or conceptual boundary that contains all the system's essential elements and effectively and completely isolates the system from its external environment except for inputs and outputs that are allowed to move across the system boundary.*

models

Mathematical representations of a system, generally capable of manipulation to simulate systems behaviour. Models are approximations to real situations, but useful in prediction, and in the development of more generally applicable theories.

input and output

Flow of materials, energy or information across a system boundary, into or out of a system.

properties

The attributes of the elements which make up a system. In the scientific use of systems theory these attributes are stated as measurements using a standard scientific system.

forcing functions

Inputs of energy or materials from outside the defined system boundary which influence system properties and behaviour.

feedback

Internal control mechanisms which influence system behaviour. Negative feedback loops tend to resist change, and thus give systems self-regulating properties.

flow pathways

Trajectory of movements of materials, energy or information. Pathways vary considerably, and in complex ways in many systems. The amounts of materials, energy and information also commonly vary over time, as the system functions.

open/closed systems

Systems, the functioning of which includes inputs and outputs (open systems) or are self-contained within the defined system boundary (closed systems). Though some ecosystems, or parts thereof, may be treated as closed systems, in reality from the terrestrial perspective all ecosystems are open, since the input of solar energy is extra-terrestrial and continuous.

black box systems

Systems, the internal structure and functioning of which are unknown or undescribed. Black boxes are useful in complex situations in which there is a hierarchy of systems. Management of biological resources may not require precise knowledge of all parts of an ecosystem. We are accustomed to using black boxes in real life. Many people have little idea of how a car works, but are able to control it well.

*Definitions marked with an asterisk (*) are quotations from Sandquist (1985). This is a good further source of information on systems concepts.

research. The systems approach has not been without critics, and there is considerable current interest in quantitative ecological research in catastrophe and chaos theories, the applications of which are a different approach to ecological problems from that of the ecosystem. Nevertheless the ecosystem concept, based on systems science concepts, remains central to most macro-scale ecological and environmental science.

Systems science is based on the principle of causality which states that a measurable cause produces a measurable effect (Sandquist 1985). In the real world the range of problems which can be investigated by systems science is very wide. Ecology and environmental science clearly belong within the category of rational knowledge, since measurement of the properties of the biological world and its environment have long been at the core of these disciplines. Systems science provides us with a powerful means of building quantitative models. Models are especially valuable in environmental science, as they allow theories to be tested. Frequently in environmental science construction of laboratory-based experiments for hypothesis testing is difficult. Models offer an alternative method of testing data. Furthermore, models may be used in prediction of outcomes of particular sets of circumstances. This may be of vital importance in environmental management. A precise definition of system, such as that given by Sandquist (1985), is rather formal. It is stated in its entirety in Box 1.4, but may be more simply summarised as 'a group of measurable elements which interact causally'. To make systems manageable a **system boundary** is defined. As with the system itself this may be an abstract concept. As far as ecosystems are concerned, these are real and tangible, and the boundaries are often defined by reference to a geographical feature or a dominant plant form, but may be defined by some conceptual human boundary, such as the limits of a nature reserve. The scope of these fundamental systems science definitions allows the ecosystem concept to be applied in many situations.

Systems and change

Systems change over time. The rate and nature of change may or may not be continuous. This change is a result of the response or output of the system by its internal actions. These are the result of system inputs which are caused by factors or stimuli from the external environment of the system. Especially in the case of very large and complex systems such as ecosystems, the inputs and outputs are complex and difficult to identify, but systems theory is sufficiently flexible to permit systems and their behaviour to be handled at a variety of levels of analysis. There are a number of major components within the system. Properties are variables in the states of the elements which constitute the system. In the case of ecosystems, this would include the characteristics of all the biota and their controlling environment at any one point in time. Forces, or more precisely **forcing factors**, are outside causal forces that drive the system. It is generally agreed in ecology that ecosystems are driven by energy, which enters the ecosystem usually as solar radiation. This supplies direct insolation to drive photosynthesis, and controls heat and moisture conditions within the biosphere, which are primary determinants of organisms' physiological processes.

Within the system, properties are linked by flows or **flow pathways**. These connect the elements and the external forcing functions through transfer of energy and materials within the system. In an ecosystem, flow pathways are the movements of assimilated energy (food energy) between different trophic levels. This must also involve flows of materials (food material), since the energy transfer is accomplished by synthesising and breaking down complex chemical compounds which carry energy in their internal chemical bonds. Interactions or interaction functions occur where forces and the system properties control flow pathways. Very important parts of most systems are **feedback** loops. These are links which take an element from a downstream part of a pathway to an up-stream location; in this way they act as control elements. In some cases the loop amplifies the output; these are termed 'positive feedback loops'. In other instances, feedback loops tend to decrease output. Negative feedback loops are as important in ecosystems as they are in both individual organisms, and in populations of organisms. Negative feedback loops act as regulatory mechanisms, tending to resist change from a steady state or equil-ibrium condition. In biological sciences these are often termed 'homoeostatic mechanisms'. Their nature and role in ecosystems remain somewhat problematic; some ecologists such as Odum contend that ecosystems possess a wide range of sophisticated self-regulation mechanisms (Odum 1971). Other ecologists have refuted this.

Systems scientists may use the terms **open** and **closed systems** to denote particular types of systems. Open systems have flows of energy or materials which pass across the defined system boundary. In the case of ecosystems, the energy subsystem is an open system. Solar radiation reaches the Earth, where some of it is used by plants in photosynthesis. This process supports most living organisms. The energy is used in metabolic activities, and is ultimately converted into heat energy which is finally radiated back to space, balancing the input of solar radiation to the biosphere. Closed systems have no movements of energy or materials across the system. An example of a closed system within ecosystems is the cycling of the majority of nutrients. Nutrients are lost from the ecosystem by movement to ocean sediments, and are gained by the breakdown of rocks. However, as the rate of such activities is relatively slow in comparison with the rate of nutrient cycling within the ecosystem boundary, nutrient cycling can be considered a closed system. For the majority of nutrient cycles, the input of nutrients from weathered rock is a minor path in terms of quantity, as well as operating at a much slower rate. Ultimately ecosystems should be regarded as open systems because the ultimate forcing factor for ecosystem function is solar radiation, and the global to local spatial patterns of variation in its supply in time. The **input** of radiation to and from the Earth is in balance, in accordance with the laws of thermodynamics, incoming solar radiation being balanced by outgoing terrestrial infra-red radiation. Within an ecosystem inputs may exceed outputs for any time scale up to the millions of years of geological time scales. In such a case some of the energy remains locked in or close to the biosphere as deposits and precipitates of organic origin. Obvious examples are coal and oil deposits. These are fossil fuels, the energy of which may be liberated rapidly by humans or remain in the deposits until broken down by natural geomorphological and geological processes over hundreds of millions of years in some instances. However, in one important respect ecosystems operate as a closed system. The supply of materials required for life, nutrients, is finite, and the cycling of these nutrients within ecosystems is essential to provide continuing support for terrestrial life. Open energy systems and closed nutrient systems are discussed in Chapters 2, 3 and 4.

Science has far to go in discovering all the detail of the function of any single organism, so such a level of understanding for ecosystems lies in the future and, indeed, may never be completely realised. However, it is perfectly possible to make use of systems, without necessarily unravelling all parts of its structure. Large systems may be broken down into a series of subsystems, the inputs to and outputs from which may be analysed without detailed knowledge of the internal functioning of the subsystem. In many instances in research this is a perfectly valid way in which to investigate the nature and behaviour of ecosystems. Most of us, living in technologically advanced societies, are used to operating (i.e. controlling) systems, the internal functioning of which we do not understand much, or even at all. Perhaps you may become a better driver if you know how a car works, but many people who are at least competent motorists have no idea of how a car functions. A system, the internal functioning of which is unknown, is termed a **black box**. The ability to use systems at different levels of analysis is most helpful in solving practical problems. Generally very big problems in rational knowledge, which require rapid solution, are best approached through systems science. This is one reason why the ecosystem concept has so much utility in biological conservation and environmental management.

Abiotic environment of ecosystems

We have established that ecosystems are complex systems of populations of organisms and their controlling environment, and that the term 'environment' includes both the abiotic or physical environment, and the biotic or biological environment. In this final section of Chapter 1 the system function characteristics of these two types of environments are outlined. The abiotic environment may be divided into a number of major subsystems, traditionally termed 'spheres'. These partially extend beyond the biosphere in some cases, and so the focus of our interest in these systems is within the 20 km thickness of the biosphere, with which all the spheres interact. This also is the most active zone of all these spheres, a fact which is related to the interaction between them. However, it should also be remembered that the subdivision of these components of the physical environment is largely for human convenience. As the biosphere and its function shows clearly, there is continuous exchange of energy and materials between all of the elements in the systems.

The atmosphere is the shell of gases around the Earth. The shell extends to thousands of kilometres above the surface of the planet, but most of this skin of gas is so diffuse as to be at near vacuum conditions by human standards. The lowest part of the atmosphere, the **troposphere**, is about 10 km thick and contains approximately two-thirds of the mass of gas which makes up the whole of the biosphere. The junction of the troposphere with the layer above, the **stratosphere**, is the tropopause, and it marks a change in the direction of the vertical temperature gradient through the atmosphere. Life is confined to the lower part of the troposphere, below about 6.5 km. Above that altitude permanent life is impossible, as the constant low temperature ensures that all water is permanently frozen. A supply of liquid water, however small and for a short period, is a prerequisite for permanent life. The gaseous composition of the troposphere is generally fairly uniform but there are exceptions to this, which though minor in volumetric terms are important for life. Box 1.5 shows average tropospheric composition. One of the most interesting properties of the atmosphere is the reciprocal relationship it has had with the biosphere since life evolved on Earth. The first life, which we would regard today as simple primitive forms, evolved in oceans which formed as the planet cooled.

Box 1.5

Gaseous composition of the troposphere

Nitrogen (N_2)	78%
Oxygen (O_2)	21%
Inert gases (mainly argon, Ar)	1%
Water vapour (H_2O)	usually < 1.0% (variable in time and space)
Carbon dioxide (CO_2)	0.035%

Box 1.6

Comparison of the Earth's atmosphere with life (now) and without life

	% With life	% Without life
Nitrogen	79	1.9
Oxygen	20	Trace
Carbon dioxide	0.04	98
Surface temperature (°C)	13	285+/−50

Source: Adapted from H.T. Odum 1983

The sub-aerial environment was hostile to life. Gradually as life evolved and developed, the composition of the atmosphere changed. Box 1.6 shows the characteristics of the atmosphere of the planet without life, in comparison to that now. The change was effected by biological action. Photosynthesis uses carbon dioxide from the atmosphere, and the reverse, oxidation process of respiration which utilises chemically stored energy returns it to the air. However, over geological time periods, carbon was effectively taken out of the rapid cycle system of the atmosphere and locked into various geological deposits in the unweathered lithosphere. Such deposits include oil, coal and limestone. Since life began, the amount of carbon dioxide has decreased until it is a very small relative component of the gaseous composition of the atmosphere. However, though small in *relative* amount, the *absolute* amount is large, and quite sufficient to sustain all current photosynthetic activity. Thus to a considerable extent the present-day atmosphere is a product of life, as well as a major life-sustaining abiotic environmental factor. Some of the implications of atmosphere–biosphere interactions, and human impact thereon, are discussed in Chapter 9.

The hydrosphere provides a second vital ingredient for life: water. Although water is commonplace, its chemistry is highly unusual. These unusual chemical properties are highly significant, both for life and its abiotic environment (Box 1.7). Autotrophic organisms (plants and some bacteria) use water in a variety of ways. It is a basic input to photosynthesis. Water is vital to the ingestion of nutrient elements, and for the movement or translocation of materials within the plant. For terrestrial plants, water plays a crucial role not only in the soil–plant root interface from which the total water supply itself is taken, but also as the only source of plant nutrients for all but a tiny handful of plants. The amount of water in the hydrosphere is large. Water exists in all states, solid, liquid and gaseous, in the hydrosphere. It is located in pools or stores which are of very different sizes. Pools are linked by flows of water, such as evaporation, transpiration, precipitation and overland flow. Some of these involve changes of state: this has great environmental significance due to the energy involved in change of state. All links are powered by heat energy derived from solar radiation. This system is called the hydrological cycle, and is shown in Figure 1.5. By far the largest store is the world's oceans comprising about 97 per cent of the total amount of water in the hydrosphere. Not only is this water unavailable to terrestrial plants due to its location, but also it is in a saline condition which only adapted marine plants can use.

Water in terrestrial environments is much scarcer, and availability of water is frequently the most important environmental condition which affects plant growth, and thus all

Box 1.7

Properties of water and their significance for ecosystems

Water is chemically and physically a substance with unusual properties. This is related to the strongly polar nature of the water molecule. These unusual properties have importance for living organisms. The main ones are outlined below.

Property	Value	Significance
Heat capacity	Only liquid ammonia is higher	Gives water a very high heat storage (specific heat) capacity. Aquatic environments have very equable thermal regimes.
Latent heat of fusion	Only liquid ammonia is higher	As liquid water turns to ice it expands. Though this is important for the vertical circulation of water, it is a major problem for living cells when subjected to freezing temperatures. Cells may rupture as cell fluids freeze and expand.
Latent heat of evaporation	Highest of all substances	Vital to water transfer in the atmosphere, and thus to the functioning of the hydrological cycle.
Dissolving power	Generally the most powerful solvent known	Vital to most metabolic processes. Examples include photosynthesis and nutrient intake by plants.

Figure 1.5 The hydrological cycle

ecosystem function in terrestrial environments. Water on land surfaces is in a variety of locations, such as groundwater, rivers and lakes, but only soil water is available to land plants, since it is through the rooting system that the vast majority of terrestrial plants take up water. Soil water is a tiny fraction of the total water in the biosphere. Plant demands on water are continuous, and indeed output of water from autotrophic plants via transpiration is a large element in productive ecosystems, generally the greater part of total **evapo-transpiration**. Therefore to enable soil water supplies to be replenished, a rapid movement from the atmospheric pool, which is also relatively small, is needed. There is a major difference in average residence period (time period that a water molecule spends in any pool) for the various pools which make up the hydrological cycle. The cycle could not function without this balance in the system. However, regional patterns of variation in rainfall, a critical element in the effect that climate has upon vegetation, is thus a major element in the abiotic environment of ecosystems. Again it is notable that the relationship is reciprocal, because plants function as an important link in the hydrological cycle, and changes in vegetation cover can have an appreciable effect upon climate at the micro-scale and in some instances at regional levels. The hydrological cycle and its significance for ecosystem function are considered more fully in Chapter 4, and the nature of climatic change, its causes and consequences are discussed in Chapter 9.

The importance of soil as a reservoir for usable water for terrestrial plants indicates one of the ways in which the lithosphere acts as an environmental control on ecosystems. The importance of the lithosphere to ecosystem function is based on priorities and functions of the topmost part, the weathered crust or regolith. Like the other spheres, the lithosphere functions and changes over time. At long time scales, measured as geological periods of millions of years, the lithosphere is subject to the processes of mega-geomorphology, such as plate tectonics. Movement of continental masses has been important in the pattern of evolution of life on Earth. Mountain building is associated with these changes. At a shorter time scale processes of erosion and deposition sculpture the detail of the surface of the Earth. These geomorphological processes are important to biosphere function in a number of ways. The cycling of nutrients is linked to geomorphological processes. Land forms provide a mosaic of different habitats, through local differences in drainage, aspect and exposure. For both the larger and smaller-scale systems of the lithosphere, a key difference with most atmospheric and hydrospheric systems is the long time scales over which they operate. Generally, lithospheric systems function over thousands and millions of years.

Soil

Strictly speaking, the biosphere includes only the very top part of the lithosphere. Soil may be considered to be the biologically active zone of the regolith. The lithosphere exerts important indirect environmental controls through the outcome of lithospheric Earth-sculpturing processes, geomorphologic actions, which shape the surface of the Earth. Surface land forms have a wide range of interactive effects, including modification of solar radiation regimes by differential surface aspect, modification of soil water conditions, and at the macro-scale control of thermal regimes through the lapse rate decrease in temperature caused by increasing elevation. The most important direct influence on ecosystem environment is through the weathering and breakdown of parent rock material to form regolith and soil. To an extent even greater than for water, the essential nutrients required for autotrophic plant growth are scarce. This acts as a fundamental control on the overall characteristics and function of many types of ecosystem, both terrestrial and aquatic.

It is not only the supply of quantities of nutrients which is profoundly influenced by lithospheric systems. The form and precise location of nutrients is crucial. For autotrophic plants to be able to use nutrients, these nutrients must be in an *available* form. This means that nutrients must be in simple ionic form in the rooting zone of plants. Although in most ecosystems cycling of nutrients through decomposition provides the majority of nutrient supply for continued plant growth, there are always losses of nutrients from the system. These losses occur because nutrients, in order to be in the available form, must be soluble. Water moving through the soil will carry away some nutrients by **leaching**. The amount of leaching which takes place will depend not only on soil water conditions, but also on overall soil characteristics, and varies throughout the biosphere. Eventually nutrients removed from the ecosystem end up in the world ocean and over geological time may be deposited as sediments, which may become sedimentary rocks. Ultimately these are broken down by weathering processes and some of the nutrients released enter the nutrient pool of ecosystems. This loop takes nutrients out of and into the ecosystem, and the time periods involved are typically tens of millions of years. Therefore this is not a part of ecosystem function, but is a part of the abiotic environmental control upon ecosystems. This process also illustrates that abiotic environmental processes involve interaction between different environmental spheres and their functional systems. Weathering and leaching of nutrients involves atmospheric breakdown of crustal materials and transport of some of the products of breakdown in solution by water moving in the hydrological cycle.

Increasingly the abiotic environment is influenced by biological action, which goes beyond the interaction between ecosystem and environment already identified. Human activities are modifying the physical environment, and thus ecosystem form and function. Direct human impact on ecosystems, such as management and replacement of natural ecosystems for agriculture, has radically changed the biosphere over the past ten thousand years or more. That this change is going on at an increasing rate is cause for global concern. This is the more so when changes to the abiotic environment, which are often accidental or unwitting in origin, cause unforeseen impacts upon ecosystems. For example, pollution is widely regarded as one of the most serious contemporary environmental issues. The essence of the pollution problem is that it causes damage to ecosystem function, and to the populations of organisms which make up ecosystems. Human beings may be among the populations directly affected, but pollution may also affect the functioning of the physical environment, and thus indirectly the functioning of ecosystems. Combustion of fossil fuels over the past century has caused an increase in atmospheric carbon dioxide content. The consequence, the so-called 'greenhouse effect', has been the first clear sign of changes in global climate. One of the most dramatic and potentially disruptive effects of global climatic change is in its effect upon world biomes. This important issue is examined more fully in Chapter 8, and yet again illustrates the way in which all environmental systems interact and influence ecosystem behaviour.

Biotic environment of ecosystems

The biotic environment of ecosystems comprises the ways in which individuals or populations of a species are affected by other members of the same species, and by members of other species, both at the same trophic level and at different trophic levels. Each individual organism is in a struggle for survival in competition with all others. This is a struggle in which there is little room for quarter, because the principle of natural selection – survival of the fittest – ensures that the very existence of the individual or even the whole species is dependent on success in competition. Issues relating to the biotic environment in ecosystems are examined in Chapters 5, 6 and 7. In this section the general characteristics of the biotic environment, and the broad ways in which these exert influences on ecosystem function are outlined.

Interactions between individuals of the same species and at the same trophic level are characterised by competition for resources for photosynthesis and nutrient inputs in the case of autotrophs, and competition for 'food' – that is, biological material with necessary content of energy and minerals – for **heterotrophs**. Individual plants will compete with other plants for light, water and nutrients from their immediate physical environment. In the case of individuals of the same species the most vigorous members will prevail over less competitive neighbours. As many species live in close proximity to neighbours, in 'clumps', this sort of competition for inputs from the physical environment is a major element in the plant's environment. Success in competition is an important factor in the continuation of the whole species survival. For consumers much the same applies except that their input, food, is previously fixed or consumed biological material. The concept of niche is highly relevant. To a greater or lesser extent all species of organisms have a specialised functional role and relationship with their environment. The niche has already been defined as the particular combination of environmental conditions which apply in that geographical location. Each species has its own particular ecological niche. No two species can occupy identical niches, though niches may seem to overlap. Niche may be considered as an organism's response to the challenges posed by competition from its neighbours, as well as a function of its physical environment.

Interactions between individuals and populations at different trophic levels relate primarily to patterns of consumption, starting with the intake of autotrophic plant tissue by primary consumers and moving up the **food chain** through secondary and tertiary consumption. Patterns of consumption involve not only herbivory and predation but also defence by plants and prey species against these actions. A further important element in consumption is consumption of dead organic matter by detrivores. The interaction by species at different trophic levels is characterised by development of the most elegant adaptational mechanisms to avoid being consumed, or to be able to capture sufficient food to survive. Organisms do not produce and store energy in tissue stores for the benefit of higher trophic levels, though feedback controls applied via consumption are vital to maintain the integrity of ecosystems. Therefore most assimilated energy is used, normally via respiration, to sustain an organism's metabolism. The ecological consequence of this is an exponential decline in the amount of energy available to support successive trophic levels. This clearly has profound implications for the character of biological communities of ecosystems. Chapter 3 deals with ecological energetics and their role in ecosystem functioning, but at this stage it is important to realise that the basis of the action of biotic environment as a controlling factor in ecosystem behaviour is energy flow and assimilation.

Competition is vitally important in regulating biological populations within ecosystems, thereby ensuring the continuing viability of the ecosystem as a whole. However, not every interaction between different organisms is aggressively competitive, though it is fair to say that the majority are. Mutualism and symbiosis are examples of collaborative interactions between different species in an ecosystem, which illustrate the complexity of biotic environmental actions. Mutualism involves an obligate relationship between two species where, for example, one species acts as a host to the other, which in turn may feed on parasites on the host. Symbiosis carries this further and the two organisms live an entirely interlinked existence. Lichens, colonial aggregates of algae and fungi, are among the best examples of symbiotic interactions. Whatever the precise nature of factors of biotic environment and their effects upon ecosystem function, it is important to remember that *all* components of the environment, abiotic and biotic, combine and interact with all the living components of the ecosystem to regulate the behaviour of that ecosystem. In the following chapters we shall be analysing aspects of ecosystem function and resultant ecosystem characteristics as individual subsystems. But we should always remember that these elements all interact in a self-regulating and unified system, the ecosystem.

Conclusions

In trying to understand the complexity of the real world, it is necessary to make abstractions and to simplify the hugely varied and changing world. If this is to have scientific validity, it must be based on measurement and testable theories. System ideas in general, and the ecosystem in particular, are a means of integrating the environment and living organisms in a scientifically sound framework. The ecosystem concept has the strength that it has evolved and developed since it was first proposed. It may be used in a variety of ways, at different scales and for different purposes. At its simplest it provides a convenient descriptive model for the functioning of organisms and their environment. At its most refined it may be used to explain the quantitative patterns of cycling of materials and energy between life and the environment. If applications of the ecosystem to particular problems have not always been wholly successful, this does not invalidate the concept. Rather it is a commentary on the ability of scientists to apply the ecosystem concept to particular problems, given current knowledge. The ecosystem concept is analysed in detail in the following chapters. It is examined in relationship to other ecological theories, and it is used to analyse relationships within the biosphere. In particular it is shown to be a most useful approach to understanding the nature and consequences of human impacts on the biosphere.

Summary

- This chapter explains the complexity of the living world and its interactions with the environment. The environment of life includes not only the physical environment of climate and so on, but also interactions between organisms of the same and of different species.
- The ecosystem concept, first used by Tansley in the 1930s, has been developed and refined since that time.
- System theory allows the ecosystem concept to be used in predictive studies and in resource management. Interactions between organisms and their environment are discussed in more detail in *Environmental Science* in this series.

Discussion questions

1   Do you think the ideas developed by Darwin have had an influence on the development of the concept of the ecosystem? If so, in what ways has Darwinian theory been important? Are there alternatives to Darwinian ideas that are useful to the study of ecosystems?

2   Draw a diagram of the structure of a small-scale ecosystem which you know. A large pond, small lake or a small wood would be suitable. You do not need to identify every species, but note the main species at each trophic level. The diagram should show the links between trophic levels, and both energy and material flows. The diagram could be assessed by a field visit, and by comparison with that produced by others for the same site.

3   Draw a diagram representing the activities of a farm growing a cereal crop, as an ecosystem. Repeat for a livestock farm. Where and how do humans fit into these ecosystems? Do this for farming systems in both more and less economically developed countries.

4   You are about to land on a planet of a solar system elsewhere in the galaxy, which appears to have a somewhat similar environment to that of Earth. You suspect that there may be some form of life on the planet. Do you think that the ecosystem would provide a useful conceptual base for the study of any life that you may discover on the planet? What fundamental information do you need to analyse ecosystems which may exist on this planet, and what problems will you encounter in obtaining this information?

Further reading

See also

Ways in which ecosystems function, Chapter 2
Energy flow in ecosystems, Chapter 3
Materials cycling in ecosystems, Chapter 4
Human impacts upon ecosystem function, Chapter 8

Further reading in Routledge Introductions to Environment Series

Environmental Science

General further reading

Basic Ecology. E.P. Odum. 1983. Saunders, Philadelphia, PA.

Odum has been one of the most powerful advocates of the ecosystem approach. Sadly, he died recently but his long contribution to ecology has been immense.

Biodiversity: An Introduction. K.J. Gaston and J.I. Spicer. 1998. Blackwell, Oxford.

A good introductory overview of this key issue.

Biodiversity. C. Lévêque and J-C. Monunoou. 2003. Wiley, Chichester.

Chapter 5, 'Biological diversity and functioning of ecological systems', sets the ecosystem concept in the context of current ideas about biodiversity. See Section 5.2 (pp. 99–100), The ecosystem approach.

Ecology (3rd edn). R.E. Ricklefs. 1990. Freeman, New York.

A comprehensive and well-written overview of contemporary ecological science.

Ecology 2. P. Colinvaux. 1993. Wiley, Chichester.

This is another excellent general ecology text, written by a distinguished scholar who has worked in both Europe and the Americas. Chapter 19 (Phytosociology) sets the ecosystem concept in the context of developing ideas about the nature of vegetation. Colinvaux is particularly good in his critical review of these ideas.

First Ecology. A. Beeby and A.-M. Brennan (2nd edn). 2004. Oxford University Press, Oxford.

Engaging, accessible and up-to-date general ecology text. Looks at ecology through hierarchy of species/populations/communities/ecosystems.

The Diversity of Life. E.O. Wilson. 1994. Penguin, Harmondsworth.

A beautifully written and scholarly yet accessible book, examining the most taxing problem facing humankind, written by one of the greatest scientists of our times.

2 How ecosystems work: operational and support functions

A brief introduction to how ecosystems work, and what they do in terms of supporting life, is needed before we examine the functioning of ecosystems in more depth. This chapter covers:

- **Operational functions of ecosystems**
- **Support functions of ecosystems**
- **Functional models of organism–environment interactions**
- **Characteristics of uninhabitable systems**
- **Trophic structure and trophic function in ecosystems**

How ecosystems work

There are two, quite distinct, aspects of how ecosystems work: their **operational functions** (that is, how the system operates) and their **support functions** (that is, what they do in terms of providing an interactive life support system for sets of living organisms). This chapter is concerned with explaining how ecosystems work, and relating operational and support functions to the concepts of system theory outlined in Chapter 1.

Operational functions of ecosystems

All ecosystems are the product of two interacting subsystems. These are an open energy subsystem (the functioning of which is described further in Chapter 3) and a more or less closed (although there are leaks along the way) cyclical materials subsystem (described in Chapter 4).

The function of the energy subsystem is to power the operation of the ecosystem. The function of the material subsystem is to provide the necessary organic and inorganic building blocks required for both the living (biotic) and non-living (abiotic) components of the ecosystem. Together the two subsystems provide for the continuing functioning of the ecosystem. If either subsystem is interrupted, degraded or altered (e.g. by pollution, or an increase in energy input, such as global warming) then the functioning of the ecosystem is likely to be altered. In turn this will affect the efficiency with which the ecosystem can perform its functions within the global environment as a whole. The efficiency of ecosystem function is important because it relates to the 'health' or resilience of the system, and thus its ability to cope with externally forced change. This includes human impacts, both deliberate through exploitation of the ecosystem, and accidental through pollution and other damaging effects (see Chapter 8).

The energy subsystem

The energy subsystem is open. This means that energy enters and leaves the ecosystem across its system boundary. The primary source of energy, which is solar electromagnetic radiation, provides light and heat needed to power ecosystem functioning. About 45 per cent of the total input of energy from the sun which reaches the surface of the Earth is in the visible wavelengths: 400–700 nm. This energy is passed through the living components of the ecosystem initially by photosynthetic fixation, which 'fixes' the energy into molecules usable by plants and other producer organisms. The energy is then passed on to **user** organisms, through consumption of plant tissue by animals (and subsequently consumption of animals by other animals) or by decomposition and organic breakdown of the resulting detritus (by fungi and bacteria).

Eventually all this energy is either locked away in the detritus (and in the past much has been locked away more permanently in organic mineral form: oil, coal, limestone) or lost as heat. During its passage through the ecosystem, the energy in living organisms is in a 'high-quality' form (at least from the point of view of the organisms concerned). It supports not only the work done by organisms in the daily activities they need to perform to survive, but also their ability to conserve and pass on the information content held in the DNA of their cells through reproduction. The concept of **exergy** is a development in ecological energetics (e.g. Jorgensen 1992) which attempts to embrace both the thermodynamics and information content of ecosystems (the latter represented by the genetic information held within its constituent organisms: an individual bacterial cell having, for example, about 600 non-repetitive genes, an algal cell about 850, a tree about 30,000 and a mammal about 140,000 genes). The operation of the energy subsystem, and recent thinking on the exergy concept as it applies to ecosystem functioning, are the subject of Chapter 3.

The materials subsystem

Life requires very specific types and amounts of materials to utilise solar energy. These materials are termed 'nutrients'. Like energy, they enter the ecosystem through autotrophic organisms. However, in contrast to the energy system, the total quantity of materials which may be used in ecosystems is strictly limited. The biosphere is a thin skin around the Earth, comprising parts of the atmosphere, hydrosphere and the surface layer of the lithosphere. Import and loss of materials to other parts of the lithosphere operate only over geological time scales of millions of years. Within the time scales over which ecosystems function, change in the total stock of some nutrients is very small. Thus the supply of nutrients which is required to support ecosystem function must be maintained by cycling within the biosphere. These cycles of nutrients are more or less closed systems, and life plays the dominant role in the cycling process. The materials subsystem is described in more detail in Chapter 4.

Support functions of ecosystems

Ecosystem function involves both the biotic activities of the living components of organisms, and the abiotic processes which go on in the non-living environment of the ecosystem. Ecosystems characteristically have a high degree of interaction between different types of functional processes. The evidence of these processes is evident in the constantly dynamic nature of ecosystems throughout the biosphere. However, to understand the way in which ecosystems function, it is important to place functional processes in a theoretical context. The primary goal of ecological science is to develop general theories which provide a consistently verifiable explanation of relationships in the real world.

The key question that must be answered is: What determines the changes in structure and species of ecosystems? The biodiversity support function is particularly crucial. Biodiversity (at its simplest, species richness, i.e. the number of species supported) is an excellent measure of the health of an ecosystem. But the potential of individual ecosystems, as we shall see later, to support sets of species depends heavily on the intensities of environmental perturbation which affect a given ecosystem. Some ecosystems have rather simple structural dynamics (especially those experiencing high-stress conditions: see Chapter 5). Others have extremely complex dynamic changes in structure and species assemblages across time or space (for example, freshwater plankton communities in lake ecosystems: see Chapter 7). All ecosystems have a hierarchy of feedback mechanisms which attempt to maintain current sets of organisms present, in the face of changing conditions. An example is the demostat feedback loops which govern population size, depending on the density of organisms present in a population (discussed later in this chapter). If conditions alter to the point where the feedback limits of these processes are exceeded, then other species are selected to replace the initial set, and a shift in species composition occurs. Some examples of such changes are described in detail in Chapter 5.

It is extremely difficult to develop models which can predict exactly which set of species, change in production, or other alteration in support function, will replace those prevailing, where an ecosystem is experiencing changing environmental conditions (particularly so given that new recombinations of genes and mutations are steadily shifting the available set of species anyway over evolutionary time). We know by empirical modelling (based on observations of real ecosystems) roughly what is likely to happen in certain cases. For example, we have quite good working models to describe the vegetation successional processes which follow disturbance of an ecosystem (see Chapter 6). Successful models usually cope with only one or a few functions of the ecosystem, over a limited range of environmental conditions. Reynolds (1996) showed that the PROTECH model can successfully simulate changes in algal abundance in a freshwater lake. This model uses five ecosystem predictor variables: water temperature, light availability, rate of flushing (water movement through the lake), availability of nutrients, and information on the grazer zooplankton present in the lake (see also Chapter 7). In other cases stochastic variability (what mathematicians call 'chaos') is too great for our existing models to cope adequately.

In order to model the likely changes in species composition and ecosystem structure associated with shifting environmental conditions, what we need are framework models which fit prevailing conditions to the known properties of sets or classes of species (rather than on a species-by-species basis). This idea underpins the rapidly developing area of functional ecology. The approach is currently based around two fundamental models of organism–environment interactions (described below).

Two theoretical approaches to understanding ecosystem function in terms of reciprocal relationships between life and its environment have achieved popularity among ecologists studying ecosystems at the community level. These are:

1   The CSR theory model, which relates plant success to the balance of stress and disturbance pressures influencing the ecosystem (Grime 1979, 2001).
2   The r–K (or 'opportunist–equilibrium') model which attempts to explain how organisms (especially animals) have developed survival strategies which best fit their ecological niche (MacArthur and Wilson 1967).

These functional models of organism–environment interactions provide a framework for understanding what ecosystems do in terms of providing a life support system for individual organisms and sets ('assemblages') of organisms. The concept of defining assemblages in functional terms provides a significant and effective theoretical base for the analysis of ecological interrelationships.

Functional groups of organisms (introduced in Chapter 1) are groups of species which show similar or analogous sets of traits for survival of a defined set of environmental conditions. Relevant traits may include morphological, physiological or life-history attributes. As a result of their shared sets of functional traits, populations of these species tend to show similar survival strategies, and to occupy the same ecosystem, or part of an ecosystem, forming characteristic assemblages. In geographically widely separated regions, which nevertheless have similar ecological characteristics, different sets of species may coexist, forming the same functional group (e.g. Hills and Murphy 1996). Various other terms are sometimes applied to this basic concept, or variations on it. For example, especially in animal species, a **guild** is a functional group of species sharing a common resource in **sympatry**, i.e. in such a way that their niches do not overlap. A famous example (MacArthur 1958) is that of warblers living in coniferous forests, where the different bird species forage for their insect food in different parts of the tree canopy. Myrtle warblers concentrate on the lower branches and forest floor, Bay-breasted warblers are mid-canopy hunters, and Cape May warblers utilise the tops of the trees. This set of bird species meets the essential criteria for functional group status. In particular, they share the same habitat conditions and food resources, and so tend to occur together within the conifer forest ecosystem.

The species forming a functional group are not necessarily closely related in taxonomic terms. Some species within a functional group may possess similar genes for the traits they have in common simply because they are closely related in phylogenetic terms (in other words they possess these genes because the species concerned share a recent common ancestor). However, very frequently functional groups are made up of organisms which are not closely related: the functionally similar attributes they possess have arisen independently in the phylogenetic history of the organisms concerned, and have been selected for by the environmental pressures unique to the particular ecosystem-type in which the functional group is found.

A good example is the isoetid functional group in freshwater aquatic plants (Farmer and Spence 1986). This group of plants is found in low-nutrient freshwater lake eco-systems worldwide (Murphy 2002: see also Box 2.1). Widespread examples of such lakes include Loch Lomond in Scotland, the Itaipu Reservoir in southern Brazil and Lake Taupo in New Zealand.

Functional models of organism–environment interactions

The CSR and r–K models examine species–environment interrelationships from somewhat different viewpoints. However, both relate 'success' to an organism's adaptations to the requirements of, and pressures on survival in, its particular niche. The CSR model in particular has proved to have applicability to a wide range of situations, allowing characterisation and explanation of the relationships which we see in communities of organisms (e.g. Grime et al. 1988; Westoby et al. 2002). The functional approach to ecology which has developed since the mid-1980s, largely based on the ideas of the CSR model, is proving to be a remarkably powerful tool for predicting how plants and other organisms respond to changes in their environment, and as the basis for practical methods for conservation of biodiversity (e.g. Hodgson 1991).

Box 2.1

Isoetids in lake vegetation: an example of a functional group of plants

Lake ecosystems with moderate to high intensities of environmental stress, produced for example by a shortage of plant growth nutrients such as phosphates (oligotrophic conditions), coupled with limitations on inorganic carbon supply in the water (associated with **acidic** conditions), have a strong tendency to support isoetid plant assemblages (Plate 1).

Plate 1 An example of an isoetid plant (*Ottelia brasiliense*) occurring in Brazilian lakes and reservoirs

Original photo: K.J. Murphy

Isoetids all look very similar, even though they come from a very wide and phylogenetically varied range of plant families. Some are closely related to the ferns (*Isoetes* spp.: which give the functional group its name). Others are dicots. Examples of these are shoreweed (*Littorella uniflora*), a member of the Plantaginaceae family, which also includes the common plantains, found as agricultural weeds throughout the world; or water lobelia (*Lobelia dortmanna*), which belongs to the Campanulaceae. Still others are monocots, such as *Eriocaulon aquaticum* (in a family of its own – the Eriocaulaceae) or *Ottelia brasiliense* (in the Hydrocharitaceae).

The characteristic morphology of these little plants consists of a rosette of leaves, often long and thin, and always arising from the base of the plant. Below the hydrosoil surface isoetids typically have a dense network of white roots, which act as pipes, taking carbon dioxide from the sediment (where there is CO_2 in abundance because of the respiratory activities of bacteria and other decomposers) and piping it up to the leaves for use in photosynthesis (thereby supplementing the limited availability of inorganic carbon from the water in their characteristically low-pH habitats). They tend to be slow-growing, often producing daughter plants vegetatively (by stolons, for example) and forming lawn-like swards on the bed of the lake.

In different parts of the world different sets of isoetid species may form the assemblage, but the functional group can nearly always be identified as present in low pH, oligotrophic lakes. Thus, for example, in nutrient-poor reservoirs in southern Brazil, species like *Ottelia brasiliense* represent the isoetid functional group. In the same type of lakes in New Zealand *Isoetes kirkii* may predominate. In oligotrophic lochs of Scotland, *Littorella uniflora*, *Isoetes echinospora*, *Isoetes lacustris* and (in the extreme west of the country) *Eriocaulon aquaticum* are the commonest members of this functional group. Across the Atlantic, in the nutrient-poor lakes of New England and western Canada, we find the stronghold of *Eriocaulon aquaticum* (the Scottish and Irish populations of this isoetid are the extreme eastern edge of its distribution, giving it a relict toehold in Europe).

Because the photosynthetic activities of green plants are the main mechanism by which the resources needed for an ecosystem to operate (especially energy and carbon) actually enter the ecosystem, plants play a vital role in determining the species composition and functioning of ecosystems. Unravelling the differences in how plants acquire, process and invest these resources may prove the key to understanding what drives ecosystem functioning. Diaz *et al.* (2004) found that easily measured sets of traits of vegetation, which occur widely, can provide useful predictors of ecosystem functioning on a worldwide basis. They showed that the sort of plant traits which 'drive' ecosystems on a global scale reflect a 'major axis of evolutionary specialization' in plant species. There is strong evidence that this represents a trade-off between the ability to acquire resources (such as light, nutrients and water) quickly and efficiently, and the ability to protect and conserve the resources once acquired within well-protected plant tissues.

In this book we have based much of our discussion and description of ecosystem support functioning upon the framework of CSR theory. The plant CSR model provides a highly effective technique for relating vegetation to its biotic and abiotic environment. Its starting point is that there are three primary sets of threats to the survival and success of primary producers (whether plants or bacteria) in the ecosystem in which they are attempting to grow:

1   **Stress** (anything adversely affecting the ability to accumulate C through chemo- or photosynthesis, i.e. pressures which *reduce productivity*, such as shade). An ecosystem experiencing intense stress conditions within all or some of its constituent habitats is likely to show low primary production: these are 'low energy' ecosystems (see Chapter 5).

2   **Disturbance** (anything which *damages or destroys the biomass* of producer organisms (plants or bacteria), either directly (e.g. grazing or forest fires), or indirectly by disturbing the habitat (e.g. unstable substrate – like a mountain scree slope)). An ecosystem experiencing intense disturbance conditions within all or some of its constituent habitats may or may not have a high primary production, but its primary producers experience a high probability of destruction of their biomass, either through biotic or abiotic causes (see Chapter 6).

3   **Competition** (effects of *other plants or bacteria*, in competitive foraging for resources such as water, light, nutrients and space). Competition is particularly important as a primary threat to survival in productive, crowded ecosystems. In these 'high energy' ecosystems organisms adapted to either high stress or high disturbance conditions are outcompeted by faster-growing, better-foraging species, and excluded from the ecosystem (see Chapter 7).

Following on from these definitions, the successful plant strategies for survival in ecosystems providing different combinations of these pressures can be categorised (see Table 2.1). Working from this simple framework, Grime (1979) went on to develop a triangular model of plant survival strategies acting in the adult (or established) phase of the plant life cycle (Figure 2.1). The features of life needed for successful occupancy

Table 2.1 Combinations of environmental stress and disturbance producing three primary response strategies in plants

| | | Intensity of stress | |
		Low	High
Intensity of	*Low* Competitors		Stress tolerators
disturbance	(C-strategists)		(S-strategists)
	High Disturbance tolerators		Uninhabitable
	(R-strategists*)		

* R-strategists are so-called because they were first identified in the roadside 'ruderal, R' habitat where trampling and other disturbance is typically high

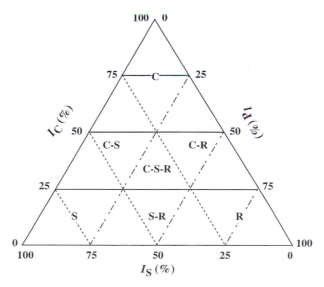

Figure 2.1 Triangular CSR model showing main and intermediate plant survival strategies in the established (adult) phase of the plant life cycle. I_d, I_s and I_c are, respectively, percentage incidence of traits for disturbance tolerance (d), stress tolerance (s) and competitiveness (c) in the plant genome. C, S and R strategists are extreme strategy types, occupying the corners of the model. Most plant species are intermediate, having a combination of traits to resist environmental pressures producing stress, or disturbance, and to forage for their required resources in the face of competition from other plants (see text and Table 2.1 for further explanation)

of the stressed, disturbed and productive compartments found in different types of ecosystems are discussed in detail in Chapters 5, 6 and 7 respectively.

But what about uninhabitable conditions, which the Grime model predicts will occur in parts of ecosystems experiencing high stress plus high disturbance conditions?

Uninhabitable systems

Certain ecosystems contain habitats which have conditions simply too hostile to plants and/or producer bacterial species (and consequently most other organisms) to allow survival. In general these systems exhibit both high stress and high disturbance (Box 2.2). An excellent example is the habitat (if we can call it that: it is virtually sterile to begin with) which remains after a major volcanic eruption, such as that of Mount St Helens in Washington State (USA) in 1980 (Plates 2a, 2b). Here an eruption estimated at about 2,500 times the power of the nuclear weapon that destroyed Hiroshima blew the top off the volcano and caused massive devastation across a large part of the upland forest ecosystem which occupied the area around the mountain. In the aftermath of this enormous disturbance event the stress resulting from the accumulation of thick layers of volcanic debris, plus a whole suite of side-effects, was sufficient to prevent any regrowth occurring for several years. Eventually however a successional process of regrowth (see Chapter 6) was initiated and by the end of the twentieth century the Mount St Helens forest ecosystem was well on the way to recovery.

A combination of high stress and high disturbance is one way to destroy or indeed prevent the occurrence of a functioning ecosystem (see Case Study 1). However, should the intensity of either stress or disturbance individually become too great, the same end result is achieved. An example is the stress conditions found in the highest mountain habitats, above 8,000 m, where reduced oxygen availability, severe cold and high intensities of mutagenic ultraviolet radiation combine to prevent the survival of most forms of life. The only life present in this essentially dysfunctional ecosystem is a few vagrants and visitors. Examples include human beings in the form of mountaineers (suffering a rather high mortality rate as the price of their attempt to enter this ecosystem) and occasional

---

Box 2.2

Uninhabitable systems

Stress *high*; disturbance *high*
↓
plants *excluded*

- Plants do not appear capable of exhibiting *simultaneously* (i.e. in same phenotype) sets of traits for tolerance of *both* high stress and high disturbance
- Examples: shifting sand-dunes (especially in very hot deserts); immediate aftermath of volcanic eruptions (new lava flows, ash falls, etc., although successional colonisation processes can commence fairly quickly); very high mountains

Plate 2 Mount St Helens, Washington State, USA (a) before the 1980 eruption showing a well-developed conifer forest ecosystem on the slopes of the volcano; (b) immediately after the 1980 eruption: the devastated and sterile remains of the forest ecosystem left after the volcano blew its top, a massive environmental disturbance event

Images with permission of James A. Ruhle & Assoc., Fullerton, California, USA

windblown insects. Even bacteria have a hard time in such extreme conditions, as evidenced by the lengthy survival time, in undecomposed form, of the corpses of those mountaineers who never made it back to Base Camp.

The r–K model of life-history strategies

This model of the survival strategies of organisms which occupy habitats within ecosystems showing differing combinations of pressures on survival divides populations of species into two groups: sometimes called 'opportunist' species and 'equilibrium' species. The terms r and K are derived from the logistic population growth model

Case study 1

An example of how the balance of stress and disturbance influences plant survival is that of ruderal weed species growing in a town in a Highland valley in Scotland. Ruderal plants (such as species of *Senecio*: Plate 3) are well adapted to the disturbance (i.e. high risk of destruction of their adult, established phase plants), associated with living in an urban townscape: like gardens and roadside verges, where human management activities pose a constant risk of destruction to the plants). These plants are common in the townscape of a skiing village like Aviemore in the Scottish Highlands. Up on the nearby cold, windswept plateau of the Cairngorm Mountains (some 1,000 m above sea level at 57°N, having features in common both with arctic ecosystems and alpine mountain ecosystems – though nowhere near as stressed as high Himalayan mountain ecosystems) lives a community of arctic-alpine plants (e.g. purple saxifrage, *Saxifraga oppositifolia*: Plate 4) – beautifully adapted to the stresses of living high in the mountains.

Certain areas of the Cairngorms are heavily disturbed by skiing developments (see also Chapter 6) and other mountain recreation activities, leading to destruction of some areas of the arctic-alpine vegetation. The problem faced by these plants is that they do not have the right set of disturbance-tolerance traits to allow them to cope with *extra* pressures on their survival when intense disturbance is added to the intense stress pressures already existing in the mountain ecosystem. Down in the valley below, however, there exist plants like ragwort (*Senecio jacobea*) which do have the right disturbance-tolerance traits: essentially they could not care less about being trampled. They also have highly mobile windblown seeds produced in vast numbers, some of which will undoubtedly be carried up to the plateau. So why do R-strategist plant species not succeed in colonising the disturbed areas of the mountain ecosystem? The answer is simple: they lack the necessary suite of stress-tolerance traits shown by arctic-alpines to survive in this hostile ecosystem. The net result is that areas of heavily trampled or heavily skied mountain slopes, throughout the world, are very quickly and easily denuded of their vegetation. The combination of high stress and high disturbance is too much to allow plants to survive.

Plate 3 A plant with a strong element of disturbance-tolerance in its survival strategy: ragwort (*Senecio jacobea*)

Original photo: K.J. Murphy

Plate 4 A plant with a strong element of stress-tolerance in its survival strategy: purple saxifrage (*Saxifraga oppositifolia*)

Original photo: K.J. Murphy

Box 2.3

Logistic population growth model

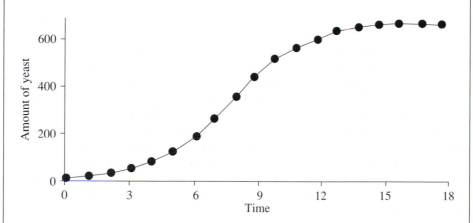

The growth curve for population increase is *logistic*:

$$\frac{dN}{dt} = rN \frac{(K - N)}{K}$$

K is carrying capacity of habitat for the population (= equilibrium level = population set point)
N is population size
t is time
r is rate of increase (birth rate minus death rate: in absence of immigration or emigration from the habitat containing the population)

(Box 2.3). They are used to describe organisms which are adapted more towards rapid production of offspring in large numbers (r-strategists) at one extreme, and those adapted to a lower rate of reproduction at the other end of the spectrum (K-strategists). Organisms which are r-selected are often quite small, are good at dispersal and rapidly fill all available parts of an ecosystem. K-selected organisms are better at competing for available resources. The individuals of a K-selected population are often larger than r-selected organisms, and tend to persist longer in the ecosystem.

In terms of understanding the population support function of ecosystems, this model certainly has its uses. However, it has a lower predictive ability than the CSR model, which allows us specifically to relate environmental pressures on survival with the prevalence (or otherwise) for groups of species adapted to different intensities of such pressures. The r–K approach is good at explaining the arrival and success of colonist species into newly opened ecosystems, and also in predicting the arrival and departure of so-called fugitive species from habitats within ecosystems (see Case Study 2 on Dutch polders).

Case study 2

In newly reclaimed Dutch polders the commonest plants to colonise the new land are often thistles (e.g. creeping thistle, *Cirsium arvense*), which produce thousands of windborne, throw-away seeds from parent populations, located on what were previously the coastal areas adjacent to the new polder. These seeds (with their parachute-like pappus to carry them on the wind) float over the polder and root into the new and empty ecosystem, forming dense populations in the first year or two after reclamation: a classic opportunist strategy. Eventually, however, they are chased out of this habitat as other more competitive species arrive, and start to fill the polder ecosystem with their own, more K-selected populations (for example scrub and tall perennial grasses, which put less effort into reproduction and more into filling the habitat with their vegetative biomass).

Although the population support function of the polder ecosystem for thistles can be explained quite well using the r-K model, the CSR model can explain what is going on at least as well, if not better. This model would describe the polder ecosystem as a newly disturbed haditat (removal of the sea water and exposure of the soil to the air creates a massive upheaval: effectively a complete change of ecosystem).

Thistles have strong disturbance-tolerance traits in their strategies, and are particularly well adapted to take advantage of newly disturbed habitats in which there are no or very few competitive pressures from other species, and conditions for growth are good (stress is low: polder soils are very fertile). Examples of these traits include production of large numbers of easily dispersed seeds, and rapid growth of plants from germination to reproduction. As long as these traits continue to give the thistles an edge over other plants in the new ecosystem, they will continue to be dominant. However, as the polder system matures, and further disturbance does not occur, then the advantages of a disturbance-tolerant strategy start to fade. In fertile soils without disturbance, species with a competitive strategy inevitably start to win the race, and eventually these more productive species (often taller to shade out the smaller disturbance-tolerators) will become dominant. However, if disturbance is maintained as a feature of the ecosystem (and in polders, which are mainly reclaimed as agricultural land, ploughing would be a frequent example of just this) then thistles are highly likely to remain as a member of the polder vegetation: these plants hang on in there as agricultural weeds.

Regulation of population size in ecosystems

Both intrinsic (e.g. social stress, territorial clashes, competition) and extrinsic factors (e.g. climate, food supply, diseases) influence the:

- abundance (size of population)
- species set (i.e. which populations, of which species, are actually present from the total set of species which could potentially occupy that ecosystem)
- rate of increase or decrease of the populations which coexist in an ecosystem

Both r–K and CSR theories make the reasonable assumption that the interactions between populations of organisms, and between populations and their environment,

Box 2.4

Demostat model of density-dependent population regulation

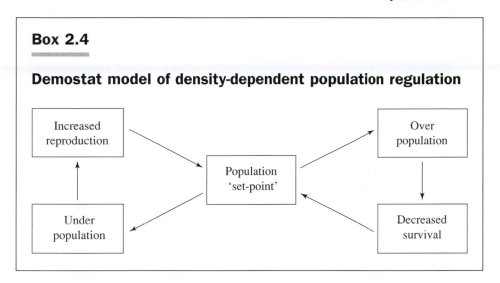

which these theories seek to model, actually occur at the level of individual organisms making up the populations.

We have already seen how population growth follows a constant, highly predictable pattern (Box 2.3). The numbers of organisms, and changes in numbers of a given population in relation to environmental pressures, are summarised by the **demostat model** (Box 2.4). This summarises how populations react to environmental pressure by feedback mechanisms, dependent on the density of the population. Thus, for example, if a population of grazing animals such as antelope increases its size to the point where the food resources start to become stretched, there will be more competition between the animals of that population for the available grazing. At the same time, the vegetation forming the food resource is becoming more sparse anyway due to the heavy grazing pressure. In such circumstances one or both of two things happens (assuming the population cannot move away to exploit a new area of the food resource, for example, by migration): either the birth rate will drop (because successful reproduction is more difficult if the animals have to spend more time looking for food, and the females are thin and less able to bear pregnancy successfully); or the death rate will increase (due to starvation, lack of milk for the young antelope, increased susceptibility to predators and disease because the animals are weak, and the likelihood that predator populations have increased simply because of the increased prey population, which is providing them with more food). The net result is a rapid reduction in size of the antelope population back towards the optimum that the ecosystem can carry for antelope. Such self-regulating feedback systems are a common feature of ecosystems, and very important for the successful long-term functioning of all ecosystems.

The actual set of populations of different species present in an ecosystem is partly governed by the tolerance ranges of individual species for each relevant environmental factor influencing survival in that ecosystem. CSR explanations of ecosystem support functions for living organisms depend heavily on this concept.

There are three identifiable sectors of the gradient of conditions which we typically see for any given environmental factor influencing the survival of a species. These are known as the *tolerant, stressed* and *intolerant* parts of the tolerance range of a given species for that particular environmental pressure. Organisms of a given species are, by definition, excluded from the intolerant sectors of the gradient. For example, if we take climate as an easily measured environmental gradient, the intolerant sectors for Thomson's gazelle (*Gazella thomsoni*), which lives in the savanna areas of eastern Africa, will be adjacent regions where conditions are much too hot and dry (e.g. the hot desert areas of the Sahara), or much too cold (e.g. the heights of high mountains such as Kilimanjaro in Tanzania) for survival to be possible. Towards the stressed extremes of the tolerance range (e.g. drier, hotter areas of savannah bordering on desert conditions), Thomson's gazelle populations will come under physiological stress. They will be thirsty, and will have to spend a lot of energy searching for water, and predators will find it easier to catch gazelle clustered near scarce water-holes. All such factors reduce the chance of survival for individual gazelles. Within the tolerant sector of their climate range (the relatively green, lush grasslands of the savanna) the gazelles experience optimum conditions for survival and reproduction. It is here that Thomson's gazelles can offer the strongest competition to other potentially competing herbivore species.

Competition usually occurs between organisms that occupy the same trophic level of the ecosystem (herbivores, C1 organisms, in the example above). Energy flow between trophic levels (introduced below) is a crucial part of ecosystem functioning.

Trophic structure and trophic function in ecosystems

The concept of trophic structure (also known as the food pyramid, trophic pathway or food web of the ecosystem) is important in understanding the operational functioning of ecosystems. Box 2.5 gives a simple example of a food chain; numerous such chains combine within the ecosystem to make up the food web (see Chapter 3). Trophic role also illustrates the importance of detrital chains within ecosystems. Without the detrital component, ecosystems could not function. This is because detrital decomposition is the dominant factor in nutrient cycling in both terrestrial and aquatic ecosystems. Chapter 4 describes the role of detritivores in the soil in more detail.

Primary producers are autotrophs: the fixers of energy in the system. All other organisms are **heterotrophs**: they use the energy fixed by autotrophs. Heterotrophs

---

Box 2.5

Trophic structure of an ecosystem: birch woodland

Organisms	Trophic level	Example
Plants	1 Primary producers (P)	Birch *Betula pendula*
Herbivores	2 Primary consumers (C1)	Peppered moth *Biston betularia*
Carnivores	3 Secondary (C2) consumers	Redstart *Phoenicurus phoenicurus*
Carnivores	4 Tertiary (C3) consumers	Sparrowhawk *Accipiter nisus*

ingest autotroph (or other heterotroph) tissues, or material derived from those tissues such as excreta and organic detritus. They then reshuffle the pack of molecules in the ingested material to suit their own respiratory and tissue-building requirements.

In the food chain described in Box 2.5 the peppered moth (*Biston betularia*) lays its eggs on the leaves of birch trees (e.g. *Betula pendula*), forming the first step in one of the energy chains leading from the producer trees to higher consumer organisms in this English deciduous forest ecosystem. The energy taken up by the feeding of the insect larvae is in turn captured by insectivorous woodland birds, such as the redstart (*Phoenicurus phoenicurus*), which feed on the larvae of the peppered moth (as well as a range of other insects). In turn the redstart is one of the food items on the favoured menu of that highly efficient woodland hunter, the sparrowhawk (*Accipiter nisus*). The final step in the transfer chain of energy held within the tissues of all these organisms is the detritus where decomposer organisms (fungi and bacteria) break down the dead tissues and excreted material produced from all these organisms, using the energy to fuel their own activities.

At every step there are thermodynamic heat losses. The exergy concept introduced earlier (and examined in more detail in Chapter 3) provides a means of modelling these functional aspects of ecosystem energetics. In Chapter 3 we also examine the classic idea of food and energy pyramids in describing ecosystem functioning, and more complex descriptions of energy transfers through the ecosystem.

Summary

- This chapter introduces the major issues concerned with the functioning of ecosystems.
- It states the theoretical frameworks of organism–environment interactions in ecosystems around which this book is based.
- It gives pointers towards the more detailed descriptions of such interactions provided in other chapters, and other books in this series.

Discussion questions

1   Are the r–K and CSR models of organism–environment interaction best suited to animals and plants respectively, or can each be applied to either type of organism?
2   Are the overarching concepts of stress and disturbance as the primary pressures on the survival of organisms in ecosystems appropriate to all organisms, some or (perhaps) none?
3   Do models such as the demostat model accurately reflect the feedback functions of ecosystems?

Further Reading

See also

System theory, Chapter 1
Ecosystem energetics, Chapter 3
Ecosystem material cycling, Chapter 4
Stressed ecosystems, Chapter 5
Disturbed ecosystems, Chapter 6
Competitive and intermediate ecosystems, Chapter 7
Human impacts on ecosystems, Chapter 8

Further reading in Routledge Introductions to Environment Series

Biodiversity and Conservation
Environmental Biology

General further reading

Plant Strategies, Vegetation Processes, and Ecosystem Properties. J.P. Grime. 2001 (2nd edn). Wiley, New York.

Provides details of the plant CSR theory introduced in this chapter. This updates Grime 1979.

The Theory of Island Biogeography. R.H. MacArthur and E.O. Wilson. 1967. Princeton University Press, Princeton, NJ.

Outlines the theory of r–K introduced in this chapter.

3 Energy flow and energetics

Fuelled by the heat and light energy of the sun, the open energy subsystem is the power-house of all but a few ecosystems on Earth. The exception is the hydrothermal vent ecosystem type, which occurs patchily along some 60,000 linear km of tectonic ridge in the world oceans, and which may provide a good model of possible extraterrestrial ecosystems elsewhere in the solar system. This chapter covers:

- **Energy inputs powering the ecosystem**
- **Ecosystems powered by chemosynthetic organisms**
- **Primary production patterns**
- **Energy flow in ecosystems**
- **The exergy concept in modelling ecosystem functioning**

Energy inputs powering the ecosystem

Solar energy is the direct driving force behind the operational functioning of nearly all ecosystems (Box 3.1). Solar energy is, on any scale appropriate to living organisms, unlimited in supply as it enters the biosphere. Input into the ecosystem is via autotrophs. Geothermal energy (Box 3.2) is a secondary source of energy which maintains the functioning especially of some specialist deep-sea ecosystems.

Approximately 45 per cent (or $c.2.5 \times 10^{24}$ J per year) of solar energy arriving at the Earth provides heat in the infra-red wavelengths (> 700 nm). Part of this energy goes to fuel atmospheric processes such as the weather machine, and part of it powers some of the cyclical processes within the material subsystem – for example, the water cycle (see Chapter 4). Much of the remainder simply warms the Earth, ensuring that most of the biosphere lies within the quite narrow range (approximately 1–30°C) which is demanded by autotrophic life on the planet. Finally thermodynamic equilibrium is maintained, by out-radiation from the Earth of the heat which is generated as a result of all of the metabolic actions of life, in accordance with the laws of thermodynamics. One of the major environmental issues of the twenty-first century, global warming, is essentially concerned with the development of an imbalance in this heat equilibrium with potentially serious consequences for ecosystem functioning (see Chapter 9).

The energy which fuels ecosystems is mainly captured by plants and photosynthetic bacteria. The sugars made by the process of photosynthesis in these organisms are the basis of the food chain in nearly all ecosystems. This process is termed 'primary production'.

Box 3.1

Solar energy supply for ecosystem functioning

In absolute terms the quantity of solar energy entering the energy subsystem at ecosystem level is about one-half of the energy of the sun reaching the top of the atmosphere. Losses are shown below:

(energy units: J per year)

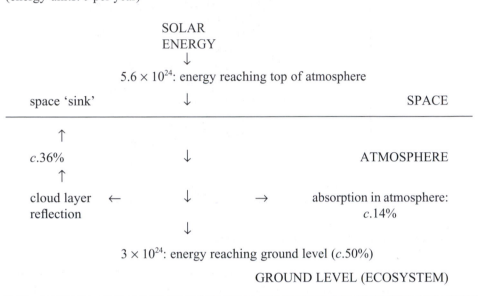

Box 3.2

Geothermal energy

Geothermal energy is derived from heat released by the earth's molten core (produced by planetary accretion processes, during the formation of the earth, and subsequently heated up by radioactive decay over several billion years). This energy becomes available at ecosystem level through volcanic activity, producing lava, hot gases, steam or other heat energy sources which may be used in the energy subsystem of certain ecosystems (such as hydrothermal vent systems in the deep oceans).

Photosynthesis and primary production

Photosynthesis is overwhelmingly the basis of primary production. It is, simply, the accumulation of energy-rich tissue by fixation of carbon to organic substances. Of secondary importance are the pathways of chemosynthesis, which occur in ecosystems where photosynthesis is impossible or severely restricted. An interesting example is

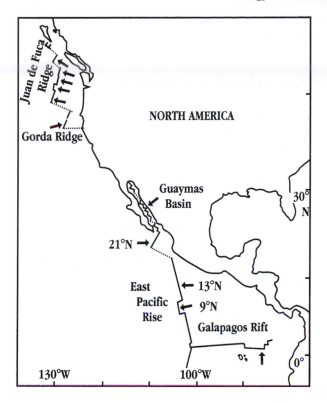

Figure 3.1 Deep-sea hydrothermal vent ecosystem sites (arrowed) in the north-east Pacific, along spreading centres (solid lines) and fracture zones (dotted lines)

provided by the deep-sea volcanic hydrothermal vent ecosystems occurring in tectonic ridge-divide regions of the world oceans. These occupy (though rather patchily) some 60,000 linear km of the deep ocean floor, for example, the so-called 'submarine ring of fire' of the northern Pacific (Figure 3.1). In one segment of this system (the Mariana Arc) a recent survey (Embley *et al.* 2004) found that of fifty submarine volcanoes surveyed, twelve showed hydrothermal activity. In addition to the oceanic systems, some deep freshwater lakes (such as Lake Baikal in Siberia: the world's deepest lake) are also now known to exhibit hydrothermal activity. Completely lacking in light, these systems are powered by the geothermal energy and minerals supplied by underwater volcanic vents, and fixed by chemosynthetic bacteria. Here bacterial production supports a functioning ecosystem type (still only poorly described: Walter 1996) which includes invertebrates such as the polychaete sulphide worm, *Paralvinella*. In turn these inverteb-rates support populations of deep-sea fish. Many of the species are still little known to biologists. Over most of the deep ocean floor only low production is possible, utilising the rain of organic debris drifting down from the sunlit surface of the sea. Hydrothermal vent ecosystems may be described as productive oases in the desert of the deep ocean floor. Such systems have been suggested as prime candidates in the search for fossil or even current life on Mars, and possibly on Europa, the volcanically active moon of Jupiter (Walter 1996). See Chapter 5 for other possible ecosystem locations on Mars.

To recap, primary production is carried out by autotrophs: plants and certain bacteria (Box 3.3). The rate at which primary production operates (i.e. the rate at which solar energy is used to convert inorganic carbon into organic substances) is the primary pro-ductivity of the ecosystem. Gross primary productivity is the total organic matter pro-duced per unit time within the ecosystem, and net primary productivity is gross primary

Box 3.3

Autotrophic organisms

Plants

Algae, bryophytes, lichens, ferns and their allies, seed-bearing plants (coniferous and flowering plants). Obtain energy for inorganic carbon assimilation (usually CO_2 but bicarbonate ions (HCO_3) in some freshwater plants) into organic compounds (sugars) from capture and use of light energy. Utilise light as their energy source and oxidise water (H_2O) to fix inorganic carbon into glucose (CH_2O):

$$\text{light}$$
$$CO_2 + 2H_2O \rightarrow (CH_2O) + H_2O + O_2$$

Photosynthetic bacteria

- Cyanobacteria (sometimes known as 'blue-green algae') which, like plants, utilise light as their energy source, and oxidise water (H_2O) to fix inorganic carbon into glucose. Common in both freshwater lakes and ocean surface water ecosystems, as well as in the surface layers of soils
- Photosynthetic sulphur and nonsulphur bacteria: again, utilise light as their energy source, but oxidise compounds other than water (e.g. hydrogen sulphide (H_2S) or organic compounds) to fix inorganic carbon into glucose. Found only in anaerobic habitats such as the surface layers of tidal mudflats:

$$\text{light}$$
$$CO_2 + 2H_2S \rightarrow (CH_2O) + H_2O + S_2$$

Chemosynthetic bacteria

Non-photosynthetic sulphur bacteria, nitrogen bacteria and hydrogen bacteria. Obtain energy for inorganic carbon assimilation by chemical oxidation of simple inorganic compounds (e.g. sulphides to sulphur). These bacteria are 'rescuers' of energy (including geothermal energy) which would otherwise be lost to normal food chains. Important in ecosystems where light is limited or absent, such as soils and deep ocean thermal vent ecosystems.

productivity minus respiratory losses of organic matter by producer organisms. This is a measure of the total 'food' available at the start of the food chain. The quantity of food which can be made available by primary producers is strongly influenced by the particular balance of environmental stress and disturbance conditions affecting that ecosystem.

Factors influencing primary production and its worldwide pattern

Patterns of primary production across the ecosystems of the world vary greatly. The lowest production is not necessarily in those ecosystems with the lowest energy inputs (hot deserts, for example, have high solar irradiance but only low productivity due to the

Table 3.1 Productive regions of the oceans

Region	Mean productivity (mg C. m^{-2}. day^{-1})
Blue waters of subtropical gyres	70
Transitions	140
Equatorial divergence and subpolar	200
Inshore waters	340
Shallow shelves	1,000

limitations on plant production imposed by water shortage). The variation in global patterns of primary production illustrates the range and variety of ecosystem character and functioning. The main factors affecting global patterns of primary production are light, heat, water, carbon dioxide and oxygen, and nutrient elements.

Occupying most of the Earth's surface, the oceans provide an excellent example of how productivity varies with regional conditions. In the oceans there are very well-mapped and predictable patterns of primary productivity (Table 3.1). Russian oceanographic research work during the 1960s (Koblentz-Mishke *et al.* 1970) defined five main productive regions of the oceans. In general, productivity is highest where there is a strong circulation of water, with upwelling currents bringing deep water to the surface. Nutrient concentrations tend to be greater in deep waters due to the continuous rain of material from shallow waters, down past the euphotic zone (where autotrophs can recycle the nutrients into the food chain). High productivity is commonest in shallow waters, where plants can grow to larger sizes.

Human impacts upon primary production are very important in many ecosystems. These are discussed more fully in Chapter 8. The broad pattern of natural environmental factors which control primary production at a global scale is shown in Table 3.2. This, however, provides only a generalised overview of ecosystem productivity (at biome level). The pattern of influence of the major limiting factors shown is frequently different at regional or local scales. Table 3.3 shows some examples of annual productivity figures for a range of plant community types in different ecosystems.

Table 3.2 Environmental controls on primary production

Type of ecosystem	Primary production levels	Major limiting factors
Oceanic (open ocean)	Very low	Macro-nutrient supply
Coastal and oceanic upwellings	Very high	Light availability below immediate surface layer
Freshwater lakes	Low–moderate	Macro-nutrient supply
Tropical rainforest	Very high	Macro-nutrient supply
Savannah grassland	Intermediate	Seasonal water supply
Desert	Very low	Perennial water supply
Tropical wetland	Very high	Macro-nutrient supply
Temperate grassland	Intermediate	Seasonal heat conditions and water supply
Boreal forest	Intermediate	Seasonal heat conditions
Tundra	Very low	Perennial heat conditions

Table 3.3 Comparative annual productivity of aquatic and terrestrial ecosystems

Type	Productivity g organic matter (ash-free dry weight) m^{-2} year^{-1}
Open ocean: dominant plants	
phytoplankton	140
Freshwater lake: dominant plants	
phytoplankon	negligible – 3,000
submerged macrophytes	
(temperate)	650
free-floating plants (tropical:	
water hyacinth)	5,000
Wetland: dominant plants	
reedswamp	
(temperate)	2,100
papyrus swamp	
(tropical)	7,500
cypress tree swamp	
(subtropical)	692 – 4,000
Terrestrial: dominant plants	
tropical rainforest	2,250
temperate coniferous (boreal) forest	900
savannah grassland (tropical)	790
temperate grassland	560

Note: Values are averages from many examples of each type
Source: Adapted from Moss 1988

While climate provides very important control factors at global scale, it is by no means the only or even the dominant control at that scale. Generally availability of water and heat act as controls when these factors are scarce in terrestrial environments. Aquatic ecosystem function is strongly dominated by the supply of nutrients. Besides providing water, aquatic ecosystems are much more buffered in terms of heat conditions than are terrestrial ecosystems. However, when there are good, year-round supplies of water on land, primary production is controlled by nutrient availability and nutrient cycling. Climate often interacts with other environmental factors in controlling patterns of primary production, and thus the whole trophic structure and function of ecosystems.

Take, for example, the way in which heat conditions affect microbial action in decomposing dead organic material within nutrient cycling systems. Nutrient transfer to plants in terrestrial ecosystems depends heavily upon the availability of soil water. Since soil water availability is obviously strongly influenced by climate, nutrient availability is in turn influenced by climate-related environmental factors. In very dry soils (such as in a hot desert) the rate of nutrient release by decomposition may be much more strongly influenced by when, and how much, rainfall occurs during a given year than by any other factor. In such conditions both plant and animal remains may persist for long periods virtually undecomposed. Along the commercial camel train ('dabuka') route from the Sudan to Egypt, through the Eastern Desert of the Sahara, which is followed by some

Box 3.4

Energy flow through an ecosystem: summary

Energy sinks		*Energy flow* $(kJ.m^{-2}.yr^{-1})$
	SUN	
atmospheric heating + weather	↓	20×10^6
	ATMOSPHERE	
water + mineral cycles	↓	4×10^6
	ECOSYSTEM	
photosynthesis losses	↓	2×10^6
	AUTOTROPHS (P)	
conversion losses (P–C1)	↓	8,000
	HETEROTROPHS (C1)	
conversion losses (C1–C2)	↓	800
	HETEROTROPHS (C2)	
conversion losses (C2–C3)		150

100,000 camels every year (Briggs *et al.* 1993), there lie thousands of mummified corpses of the camels which died *en route*. Rainfall on this part of the planet's surface occurs only about once every seven years. The full decomposition of the remains of the dead camels is an exceptionally slow process.

Consumption, predation and decomposition: energy flow in ecosystems

The energy captured by the ecosystem is transferred through different levels of the trophic structure of an ecosystem by consumption and predation. Ultimately all biological energy is converted to heat via respiration. However, this conversion can be postponed when energy is stored in the form of biodeposits (e.g. coal, oil). Decomposition processes, in which ecosystem detritus is broken down by the micro- and macro-organisms which have specialised in the consumption of dead organic matter, play a crucial role in ecosystem functioning. The detrital subsystem illustrates well the interaction between energy and material systems.

The starting point for the flow of energy through the ecosystem is usually sunlight. The subsequent efficiency of energy capture and transfer through succeeding trophic levels is shown in Box 3.4.

Energy pyramids, food webs and stable isotopes

Most of the available energy not lost as heat is used, at each trophic level of an ecosystem, to support the operation of metabolic pathways within the organisms dominating that level. Pyramid diagrams can be constructed to show the amount of energy (or biomass, or numbers of organisms, which are approximate indicators of energy content) tied up in the biota of the ecosystem at each level (see Figures 3.2a–c). These relationships are important in appreciating the spatial and temporal patterns of distribution of organisms in ecosystems.

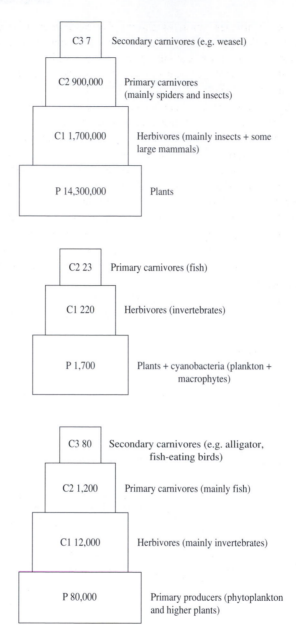

Figure 3.2 Pyramid diagrams depicting trophic relationships in ecosystems. See Box 3.4 for explanation of P, C1, C2 and C3 trophic levels (a) pyramids of individuals per ha in a Kentucky bluegrass grassland ecosystem; (b) pyramid of biomass (kg per ha) in a Wisconsin lake freshwater ecosystem; (c) pyramid of energy: energy flow (kJ per m^2 per year) through Silver Springs, Florida: a wetland ecosystem

For example, this approach may be used to model how much energy there is available, in a given ecosystem, to support producers and consumers at different places and times.

Pyramid models illustrate clearly how successive levels within the food chain (from producers to herbivores, to carnivores) support each other. They give a clear indication of 'why big fierce animals are rare' (Colinvaux 1980). Such animals are at the top of the energy pyramid, and there is simply not enough energy available up there to support a large biomass, or number, of top-level carnivores. This is especially so for warm-blooded

homiothermic carnivores, such as tigers (*Panthera tigris*) or orcas (killer whales: *Orcinus orca*), which have a very high energy demand compared with cold-blooded **poikilothermic** carnivorous animals such as crocodiles (e.g. the Nile crocodile, *Crocodylus niloticus*) or sharks (see Chapter 5). All else being equal (though in practice it never is) we would expect to see a marine ecosystem being able to support a higher biomass of great white sharks (*Carcharodon carcharias*) than of orcas, simply because the sharks require less food to enable them to function, and so the ecosystem can potentially support more of them. In reality this simplistic view ignores a large number of other factors which are important in determining the relative success of homiothermic and poikilothermic carnivores (such as behavioural factors, and competitive ability). Most ecologists would agree that the orca is substantially more successful than the great white shark as a top-level carnivore in the world's oceans.

This example illustrates the limitations of models which attempt to predict ecosystem functioning solely or largely on the basis of energy flows. Such models can certainly identify the functional limits of an ecosystem for supporting biomass or numbers of organisms at each trophic level. But this is at best only a rather crude overview of what is actually happening.

In reality, the trophic structure of an ecosystem is a complex web of food and energy flow relationships known as food webs. Figure 3.3 gives an example of a food web for

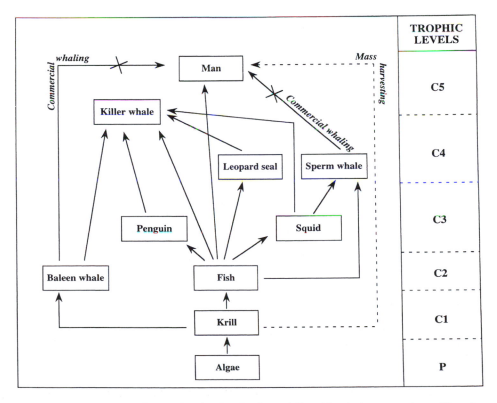

Figure 3.3 Antarctic Ocean food web, showing feeding relationships between producer (P) and consumer (C1: herbivore; C2–5) organisms. Consumption of whale meat by human beings is now only a small fraction of what it used to be, but there are pressures for the resumption of whaling. Mass harvesting of krill for human consumption remains a small fishery, partly because of heavy metal contents in krill. However, this fishery is likely to increase in the future. Note the complexity of trophic links for top-level carnivores (C5) such as killer whales (orca), which feed on a variety of organisms from the three next lowest trophic levels (C2–C4)

the Antarctic Ocean: in one of the more productive regions of the seas. Note how the trophic relationships become more complex at higher levels: top carnivores such as the orca may feed on a wide variety of lower trophic-level organisms. A traditionally depicted aquatic food web of this type also ignores a potentially important aspect of production, via the **microbial loop**. This is a heterotrophic bacteria-based pathway which, ecologists are now discovering, is important in many if not all aquatic systems (and potentially important in at least some terrestrial soils as well: Bonkowski 2004). It functions via microbial organisms, such as ciliate protozoa, which consume bacteria living on particulate detritus and other forms of organic matter within the water column, and in turn provide a food supply for organisms higher up the food chain (e.g. copepods) additional to that provided by the primary production of cyanobacteria, algae and other aquatic plants.

Cascade theories of ecosystem regulation suggest that the consumer organisms (including both top carnivores, such as sharks or orcas, and lower-level herbivores: C1 animals, such as krill in the Antarctic Ocean food web) present in ecosystems may have an important regulatory effect on producer organisms (Cohen *et al*. 1990). The 'top-down' model of community structure implies that predation has an important effect on food webs, though modified by 'bottom-up' influences of nutrient availability. Overgrazing of grassland or shrubland ecosystems is a classic example, discussed in Chapter 8 (e.g. McQueen *et al*. 1989).

Evidence from lake studies (see Case Study 3) further suggests that such models can be quite successful in explaining interrelationships between organisms occupying different trophic levels in ecosystems.

Case study 3

In the Norfolk Broads of southern England (shallow lakes produced by flooding of areas left after removal of peat (mainly for fuel) in the Middle Ages) cascade theory has been applied for practical management purposes, to reverse problems of **eutrophication**. These lakes have high nutrient loadings, derived from fertiliser runoff from surrounding agricultural land, and sewage inputs from towns and villages in the area. During the mid-twentieth century the once-clear water of the Broads became turbid as nutrient pollution encouraged the growth of large phytoplankton blooms (see also Chapter 7) in the water, producing foul-smelling green soupy water. Part of the solution to the problem entailed better water treatment to reduce the nutrient loading from the surrounding catchment. However, the rate at which the lakes recovered was greatly improved by manipulating the consumer organisms of the lake ecosystems.

As in many freshwater lakes (e.g. Figure 3.3b) the primary consumer (C1) organisms were various species of zooplankton. The population density of these little herbivores depended largely on the quantity of their food source (the algal cells of the phytoplankton), but was also heavily regulated by the numbers of fish (C2 organisms) predating the zooplankton. By reducing fish numbers, through active management measures, in the lakes, it proved possible to reduce the predation pressure acting on the C1 level. The result was an explosion in numbers of zooplankton organisms, which rapidly consumed the algal blooms. The outcome of this top-down management was a substantial reduction in algal bloom problems. This happened much more quickly than would have been expected if only the nutrient supply had been manipulated, by bottom-up management of the sources of nutrients entering the lakes.

Cascade models help explain how the all-important producer organisms (important, of course, because they define how much energy will enter the ecosystem at the base of the energy pyramid) may respond to environmental influences produced by predation (either direct grazing pressure, or indirectly by the influence of carnivores on the populations of grazing animals). Quite commonly, however, the rate of primary production is not seriously affected by the variations in predation intensity which may be going on at and between higher trophic levels in the ecosystem. When this is the case, it is difficult for a cascade model to predict changes in ecosystem functioning with much success.

An approach which can provide us with an appropriate tool for tracking resource movement through food webs is stable isotope analysis.

The atoms which make up an element must, by definition, all have the same atomic number (i.e. number of protons in their nuclei), but are not necessarily all of the same atomic weight. The small differences in atomic weight are produced by differences in the number of neutrons present in the nucleus. Atoms of an element with the same atomic number but different atomic weights are **isotopes**. Some of these are unstable and radioactive. However, most elements which are of biological significance have at least two stable isotopes, the lightest of these being much commoner than the others. The heavy isotopes of carbon and nitrogen (^{13}C and ^{15}N) are present in very small and varying quantities in all ecosystems (about 1.11 per cent of the carbon, and 0.37 per cent of the nitrogen, on average, for terrestrial ecosystems). These isotopes are proving to be extremely useful tracers of the movement of C and N through food webs. This is because their quantities (measured using Isotope Ratio Mass Spectrometry: IRMS) in tiny samples, usually only a few milligrams, taken from different parts of the environment, or from living organisms, provide very good ecological signatures (usually written as $d^{13}C$ and $d^{15}N$, and measured in ‰: 'parts per thousand') of the source of the C or N in the tissues of a given organism.

The flow of nutrients through food webs in ecosystems supporting very high biodiversity is notoriously difficult to tease out using traditional methods. For example, the riverine wetlands of southern Brazil, such as the Upper Rio Paraná, have very complex food webs, supporting large numbers of species. In the Upper Rio Paraná wetlands (Agostinho *et al.* 2000), the numbers of taxa (not all have yet been identified to species level) so far discovered include:

- Phytoplankton: 300
- Zooplankton: 329
- Periphytic algae: 228
- Bentic macroinvertebrates: 80
- Aquatic macrophytes: 48
- Non-aquatic plants: 450
- Fish: 170
- Amphibia: 22
- Reptiles: 37
- Birds: 256
- Mammals: 60

To give some idea of how stable isotope analysis can help unravel the complexities of nutrient and energy flow through highly diverse systems such as this, consider what happens at the bottom of the food chain. In plants, the process of photosynthesis results in strong fractionation of carbon isotopes, depending on the type of photosynthetic pathways operating in the plant. Most plants use the so-called C3 pathway (which converts CO_2 to a phosphoglycerate compound with three C atoms). On the other hand, some warm-climate plants, especially grasses (such as the tall emergent grasses which dominate Brazilian floodplain wetlands), have evolved a different type of photosynthesis, the C4 pathway (which converts CO_2 to dicarboxylic acid, with four C atoms, and is more efficient in hot, sunny conditions). C4 plants have higher $d^{13}C$ values ranging from −17‰ to −9‰, while the range of $d^{13}C$ for C3 plants is −32‰ to −20‰. Detritivorous fish using the floodplain water bodies as their feeding ground would potentially be able to feed on detritus derived from either the C4 swamp grasses, or from a range of C3 species growing in the surrounding forests. Once incorporated into the tissues of the fish, the signature of the carbon is preserved, and can be picked up higher in the food chain by analysing, for example, muscle tissue of carnivorous fish such as piranha (*Serrasalmus* spp.) to find out where the carbon that fuels the wetland ecosystem is primarily coming from (either **allochthonous** inputs, such as tree leaf litter; or **autochthonous** sources, such as the swamp grasses within the wetland itself). This sort of information is potentially of practical importance in managing the wetlands to preserve their biodiversity. If we know the relative importance of different sources of carbon for the organisms in the ecosystem we are much better able to make properly informed decisions about how best to manage the system to preserve those carbon sources (i.e. plants) in the correct proportions and quantities to maintain the optimal biodiversity support functioning of the system.

The exergy concept

The exergy concept (Jorgensen 1992) is the basis of current attempts to model organism–energy interactions in ecosystems. This approach aims to develop thermodynamically based models of ecosystem functioning, which may, for example, allow us to predict how the biota of an ecosystem might respond to specific environmental changes. Exergy is *To (I)*, where *To* is the temperature of the environment and *I* is a measure of the 'thermodynamic information' of the system. It is effectively a measure of how far above the thermodynamic equilibrium (the state at which a system containing no living organisms would exist) the ecosystem is operating.

The more living matter there is in an ecosystem (and the more complex that living matter is in terms of its genetic information content), the higher its exergy (see Chapter 2). The exergy concept may be able to explain and predict the success of sets of species in a given set of ecosystem conditions. 'Living organisms develop and evolve in ecosystems due to the throughflow of energy (exergy) . . . the combinations of organisms with the properties that ensure best the maintenance of biomass, and their embedded [genetic] information under the prevailing conditions, will be the survivors' (Jorgensen 1996). Such organisms will have the highest exergy within the system. This is effectively a restatement of Darwin's famous 'survival of the fittest' theory, but is couched in energy and biological information content terms.

On a practical basis it is possible to calculate exergy, at least approximately, if we know the genetic information content (*Wi*) of each of i components of the system (from organic detritus – which has zero information content, to mammals which have a very high information content: see Chapter 2), the biomass (including that of once-living organic detritus) of each of these components, and the volume of the ecosystem in which the exergy is held. The units are g detritus-equivalent per litre. To convert to energy units is a simple matter since on average the amount of work energy which 1 g of detritus can do is 18.8 kJ. So we end up with an estimate of exergy in kJ/litre. Exergy can be calculated for individual components (such as a species or sets of species) or for the ecosystem as a whole.

Figure 3.4 Plot of energy v. Si/P ratio for two diatoms with different half-saturation constants for Si and P respectively of P: 0.003 and Si 0.5 mg l^{-1} (curve marked ■) and P: 0.1 and Si 0.5 mg l^{-1} (curve marked o). See text for interpretation

Some examples of how this concept may be used in practice to model species–environment interactions are given by Reynolds (1996). In lake ecosystems the ratio of phosphorus:silica (Si/P) is important in selecting the 'winners' of competing diatom species within the phytoplankton (see also Chapter 7). If the ratio is high, diatoms with a low half-saturation constant for P are selected (these are species which can grow well on short rations of the essential nutrient P: see Chapter 4 for more on nutrients). If the ratio is low, species with a low Si half-saturation constant tend to win the race for dominance. Figure 3.4 shows a plot of exergy for two species of diatoms with contrasting half-saturation constants for these nutrients across a range of Si/P values typical of freshwater lakes. The point at which the curves cross corresponds very closely to the observed point on the Si/P gradient where a shift in dominance between species with these differing adaptations actually occurs in reality (Tilman and Kilham 1976). This sort of evidence suggests that the idea of using 'maximum exergy' as a predictor of species success, in modelling ecosystems, may well be worth further investigation.

Summary

- This chapter summarises the principal features of the energy subsystem within ecosystems, and shows how this supports the trophic structure and trophic functioning of ecosystems.
- It outlines the importance of primary production, by plants and certain bacteria, as well as bacterial production in some types of ecosystem, as the basis for flows of energy through food chains and more complex food webs, and discusses methods for analysing such resource fluxes. The concept of exergy is discussed.
- More detailed accounts of ecosystem energetics are provided in two other books in this series: *Energy Resources* and *Environmental Chemistry and Physics*.

Discussion questions

1 How best can the flow of energy through an ecosystem be represented?
2 Photosynthesis and bacterial chemosynthesis are the most important mechanisms of primary production that we know about. Might there be others, and if so where should we search for the ecosystems which may contain such 'different' producers?
3 How might recently developed ideas on the energy functioning of ecosystems, such as the exergy concept, be used by ecologists interested in modelling ecosystem responses to environmental change?

Further Reading

See also

Ecosystem energetics, Chapter 3
Ecosystem material cycling, Chapter 4
Low energy ecosystems, Chapter 5
Productive ecosystems, Chapter 7
Grazing, Chapter 8
Heat equilibrium, Chapter 9

Further reading in Routledge Introductions to Environment Series

Energy Resources
Environmental Chemistry and Physics
Energy, Society and Environment

General further reading

Ecology 2. P. Colinvaux. 1993. Wiley, New York.

A highly regarded general ecology text that gives a clear and critical discussion of energy flow in ecosystems.

4 Material cycles in ecosystems

The materials which make up ecosystems, nutrients, are the building blocks of life. Nutrients enter ecosystems through the metabolism of autotrophic plants. These organisms have specific requirements in respect of type, amount, location and form of nutrients. The latter two requirements are related to the concept of nutrient availability. Within ecosystems, cycling of nutrients, mainly by soil organisms in terrestrial environments, is vital because nutrients are in limited supply. Nutrient cycles function as closed systems. This chapter covers:

- **Nutrient availability**
- **Nutrient cycles**
- **Soil, nutrient cycling and nutrient store**
- **Types of nutrient cycling systems**

Materials: the building blocks of ecosystems

The way in which ecosystems use energy to power their functioning was analysed in Chapter 3. The transmission of energy through an ecosystem is dependent on the availability of specific materials. A central feature of all material used in ecosystems is cycling. Without cycling, ecosystem functioning would rapidly come to a halt. This chapter is concerned not only with what materials are involved in ecosystem function, and their specific functional roles, but also the ways in which different ecosystem materials are constantly cycled within the biosphere. General types of cycling systems which are based on particular nutrient elements can be defined. However, the detailed pattern of nutrient use and cycling within an ecosystem depends on the specific character of that ecosystem, in particular the nature of autotrophic vegetation and primary production in that ecosystem, and the characteristics of the physical environment. In the analysis of material cycles in ecosystems the biological focus will be on two categories of organisms: autotrophs and detritivores. The former are responsible for the intake of mineral nutrients into the ecosystem, and thus are the starting point for virtually all the material flow in the living components of the ecosystem, while the latter break down organic tissue, returning partly or wholly disaggregated material to the soil in the case of terrestrial environments. The former use nutrients to construct the substance of life; the latter are a major factor in the return of living material to a simple abiotic form which may be used again by plants.

The term 'nutrient' is used to describe the chemical elements which are used to construct living material. This needs to be explained more fully. The input of materials into the ecosystem, as with the energy input, commences with autotrophic plants. This is a logical starting point, because the input materials are in their simplest chemical combination. However, it should be remembered that material use in an ecosystem occurs in a continuous and largely closed cycling system. Not all the chemical elements which exist on Earth are involved in the construction of biological materials, or at least not in quantities which have, as yet, been detected. The majority of material, typically at least 90 per cent of total **biomass**, is composed of compounds of three elements (carbon, hydrogen and oxygen), the so-called major nutrients. These materials are derived from ingested water and carbon dioxide, either directly from the atmosphere or from air dissolved in water. The remainder of the content involves fewer than thirty elements, in any measurable quantities. Box 4.1 lists the nutrients by type (major nutrients, macronutrients and micro-nutrients) and by their proportions in the biosphere. These nutrients are taken in, in solution from the soil, or from the atmosphere in gaseous form for terrestrial plants; or from the surrounding water in the case of aquatic plants, and from the hydrosoil in the case of rooted aquatic plants and photosynthetic bacteria. These elements are divided into two groups according to the amounts used in plants. Eight of them are macro-nutrients, which are elements generally required by plants in quantities measurable in parts per hundred or per thousand. Micro-nutrients, which comprise the second group, are required in very small quantities. In some cases this can be as little as a few parts per million of the total biomass of the plant.

However small the amount of nutrient required, it is, none the less, essential for plant growth. Figure 4.1 shows that plant growth response to variation in nutrient supply is a humped curve. The exact relationship varies, with plant species and also for individual nutrients, but in general there are three crucial points. There is an optimum, which is the supply of nutrient which is ideal for a particular plant–nutrient combination. Decreasing or increasing the supply of nutrient will cause a decrease in growth rate. This is probably due to increasing physiological stress as supply decreases below the optimum, and due either to stress caused by a toxic response to the high presence of the element, or to competition by faster-growing species in the nutrient-rich part of the supply curve (see Chapters 5 and 7 for more on life in stressed and competitive ecosystem conditions). There are also two points beyond which growth will cease: a minimum, below which growth does not occur, and a maximum, again beyond which growth does not occur.

This situation may be likened to trying to construct a complex model building from a set of instructions, using children's interlocking bricks. The building must be constructed from a precise mixture of different numbers and shapes of brick. If there are not enough of some sorts of bricks then the building has to be scaled down or some feature left incomplete. Extra bricks cannot be used, and in sufficiently large excesses may hinder construction to the extent that it slows and stops. A highly depleted supply of bricks means that no building is possible. In this illustration the various sorts of bricks represent specific nutrients. Some – such as the standard rectangular bricks – are needed in large amounts. Others – such as bricks for windows – are needed in very much smaller amounts, while those bricks which represent fittings (e.g. furniture) are needed in only ones or twos. This is like major, macro- and micro-nutrient supply requirements. However, the complete building needs all of these in the specified amounts; otherwise a complete and perfect replica of the plan cannot be constructed. Plants behave rather like this, but in a much more complex and elegant way. Our 'building blocks' are nutrients, which autotrophic plants ingest from their abiotic environment. The 'plan' is the genetic code for that plant, contained in its DNA. The optimum supply of nutrients will allow maximum growth, assuming that there are no other environmental constraints. Too much or too little of a nutrient will result in a toxic or sterile environment, inhibiting growth.

The supply of nutrients from the abiotic environment varies continuously in time and in space, and is thus a critical, though not the only, determinant of the amount of autotrophic plant activity (i.e. primary production) in a given ecosystem. Factors influencing patterns of primary production, and the resultant global pattern of biomes,

Box 4.1

Major, macro- and micro-nutrients, showing the relative proportions of each element in the biosphere

Element		Percentage composition in the biosphere
Major nutrients		
Hydrogen	H	49.7
Oxygen	O	24.9
Carbon	C	24.8
Total		*99.4*
Macro-nutrients		
Nitrogen	N	0.27
Calcium	Ca	0.073
Potassium	K	0.046
Silicon	Si	0.033
Magnesium	Mg	0.031
Phosphorus	P	0.030
Sulphur	S	0.017
Aluminium	Al	0.016
Total		*0.516*
Micro-nutrients or trace elements		each < 0.001
Sodium	Na	
Iron	Fe	
Chlorine	Cl	
Fluorine	F	
Iodine	I	
Manganese	Mn	
Cobalt	Co	
Copper	Cu	
Zinc	Zn	
Vanadium	Va	
Tin	Sn	
and others at very low concentrations		
Total		*< 0.01*

Source: Data adapted from Deevey (1970)

are analysed in Chapter 8. The nutritional challenges imposed by the environment on an individual ecosystem strongly influence the type of vegetation which can occur there: only plant species with the appropriate survival strategy to meet these challenges can flourish. This applies to both aquatic and terrestrial ecosystems, and is an important determinant of both the spatial pattern of vegetation on Earth, and the functioning of ecosystems within the biosphere. Functional ecology in general, and plant strategy

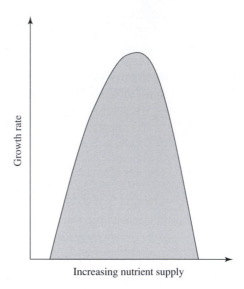

Figure 4.1 Relationship between plant growth and nutrient supply

theory, in particular improve our understanding of the ways in which plants cope with the challenge of their nutrient environment.

Control of primary production by variation in the supply of essential nutrients is largely governed by the relative supply of each individual nutrient. This is known as Liebig's Law, or the Principle of the Limiting Factor. Liebig first proposed the theory in 1840. Although subsequent research has shown that resource factors, including nutrient supply, can act in an interactive way, the 'Law' gives a good indication of the dependence of autotrophic plants on the supply of nutrients. Other environmental factors are also of importance. For example, if supplies of nutrients and water are abundant, plant growth will be slow or non-existent if temperature conditions are below the threshold for growth. It must be remembered in the following discussion of individual nutrients, and their cycling within ecosystems, that these are subsystems within the whole ecosystem structure.

In the latter part of this chapter, two major nutrients, two macro-nutrients and their cycle systems are analysed in detail. In each case the general approach to analysis has broadly the same structure. The two macro-nutrient cycles have been selected to illustrate two types of nutrient cycle systems. These have relevance to both macro- and micro-nutrient cycles. The two cycle systems contrast in several ways. The examples show that there is interaction with all the realms or spheres which interface with the biosphere, but the character of each of the examples is dominated by particular processes that take place in the hydrosphere, the biosphere, the atmosphere or the lithosphere. To appreciate the distinctive nature of each cycling system, we need first to examine some general issues relating to nutrients and their cycling, and to briefly examine the role of soil in the nutrient cycles of terrestrial ecosystems.

Nutrient cycles and soil stores: the concept of availability

At the most basic level, all nutrient cycles have the same structure. A very simple statement of this is given in Figure 4.2. This shows that material inputs into ecosystems exist in two forms – available and unavailable. The concept of availability has two dimensions. To be available a nutrient must be in a particular location and it must also be in a

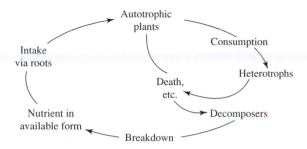

Figure 4.2 Generalised nutrient cycle system. For terrestrial plants, available form means in solution in the rooting zone of the clay and humus soil colloids (rhizosphere). Soil colloids act as reservoirs holding available nutrients so that they can be taken in by plants

particular form. If they are to be usable by plants, nutrients must be accessible to plants' mechanisms of ingestion. For most terrestrial plants this means immediately adjacent to the active parts of plant rooting systems. The only exception to this is carbon, which is taken in through plant leaves directly from the atmosphere in the form of atmospheric carbon dioxide. In the case of aquatic plants it means the water, or more properly solution, surrounding the plant's roots and foliage. In both cases the nutrient has to be in a simple ionic, water-soluble form. This means that water in the soil is vital for the nutrition of terrestrial plants, as well as having an equally vital role as an input to photosynthesis. One of the problems plants face in obtaining material supplies is that in all cases except that of carbon dioxide in air, the nutrient in solution is in a potentially highly mobile condition. This is because water, together with its solute content, tends to move rapidly downwards in the soil, away from plant roots due to gravity, or in aquatic environments can be carried out of reach of the plants by water movements. In both cases it may be readily rendered unavailable by chemical precipitation. Biological and chemical processes within nutrient cycling are vital in constantly replenishing the available pool.

Unavailability may also result from conditions other than transport to some part of the abiotic environment remote from the plant, or conversion of the nutrient element into an insoluble chemical form. In many ecosystems the bulk of nutrient supply is unavailable because it is a component of biological tissue, either as part of a living organism, or an organic residue, such as dead plant material. Most organic material is not soluble, so that after it enters the detrital food chain, the most important change to material is decomposition. This is accomplished by a vast range of biota, from bacteria to macroscopic invertebrates. Many of these are specially adapted to digest, and thus break down the materials most resistant to biochemical change. This is a significant component of most residues. Some breakdown processes are carried out by chemical and biochemical action, such as oxidation or hydrolysis. In reality the breakdown of organic material is complex and varies considerably. The particular breakdown path depends on the type and quantity of input of organic material. Breakdown, the return leg of the general nutrient cycle, is thus carried out by an interactive complex of biological and chemical actions. These mainly take place in the soil (or hydrosoil of aquatic ecosystems). The functioning of the soil component of ecosystems is of fundamental significance to the whole of ecosystem functioning.

In aquatic environments, nutrient deficiency is the most common limiting factor for primary production. In water bodies of all but the smallest spatial dimensions, nutrient cycling is made more difficult by the tendency of soluble nutrient minerals to be removed from the surface zone of water bodies, where light is available for photosynthesis, to deeper, darker locations, which cannot sustain autotrophs. Nutrients may become locked in these locations, forming chemically precipitated and particulate sediments, the nutrient content of which remains unavailable for geological time periods. In many aquatic ecosystems, not only is nutrient supply very limited, but also the cycling links are fragile and easily disrupted. In some aquatic ecosystems, damage to vulnerable biological or physical cycling links results in increased nutrient deficiency, and thus impaired ecosystem function. In other aquatic ecosystems, nutrient cycling problems are related to sudden surges of particular nutrients caused by pollution of human origin. The resultant increased nutrient availability or **eutrophication** may result in explosions of primary production, and consequent disastrous changes in the overall life support ability of the water body (see also Chapters 7, 9 and 10).

Soil and nutrient stores

There are general issues relating to the ways in which the soil component of the ecosystem functions, and the controls upon this function. The outcome, of course, is the enormous variety of soils, which matches the range of types of ecosystems over the Earth's continents. The key general factors influencing the working of the detrital component of ecosystems are the nature of the input of organic debris, the soil community and the soil environment. Organic detritus includes the remains of dead plants, animals and micro-organisms, and plant and animal exudates and excreta. It is because everything living ends up eventually as organic detritus that this material is the unit used in calculations of exergy in an ecosystem (as discussed in Chapter 3). However, it is plant debris or litter which makes up the overwhelming mass of input material in most ecosystems (typically 90 per cent), and it is the nature of the contribution of plant litter which shapes the starting point of decomposition. Much plant tissue is quite resistant to biological or chemical change. Few herbivores can extract more than 50 per cent of the available energy in food which they consume, by digestion. Thus a significant amount of plant litter reaches the soil unaffected by consumption. Lignified material – wood – is particularly resistant to breakdown, as the durability of furniture made from wood demonstrates. The energy locked in this material is the food store for soil biota. There is a vast array of these organisms, including micro-organisms such as bacteria and protozoa, heterotrophs such as fungi, and macroscopic animals such as ants, termites and earthworms. These primary decomposers in turn are the prey for predators such as spiders, centipedes and a range of larger secondary consumers including birds and soil mammals such as moles. At the top end of the food chain there are significant linkages with the surface food chain, with higher carnivores predating animals from both the autotrophic and the detrital systems. As well as chemical decomposition, the functioning of the detrital food chain physically mixes and moves decomposed material. This is another critical element in nutrient cycling satisfying the locational criterion in the availability concept. To be available, nutrients must be where autotrophic plants can get at them.

Soil organisms are among the most specialised of all biota. The food requirements of most soil consumers is specific. Thus the nature of input debris is very important. Some types of material are much more readily broken down than others. Cellulose is a singularly resistant substance. However, some organisms have developed potent capabilities to decompose such material. Among such, termites, ants and a range of bacteria are preeminent. In physical environments in which these creatures can flourish, wood is rapidly and completely broken down. Leaves and other green matter are much less resistant to breakdown. In temperate climates deciduous trees contribute an annual 'rain' of litter from autumn leaf-fall. In these conditions bacteria and earthworms are efficient decomposers, but being poikilothermic, or 'cold-blooded', there is also a marked seasonal pattern in rates of organic matter breakdown. The mixing action of worms is a major element in the nutrient cycles of the soils of such areas, by which means available nutrients, the product of decomposition processes, are spread through the rooting zone.

The physical environment of the soil (particularly soil climate, heat and water conditions) is a major control on decomposition of organic matter. The effect of seasonality as a climatic control on the metabolism of detrital feeders has been noted already. Persistent temperatures below the range required for plant growth or poikilothermic metabolism, or lack of water input from rainfall to allow significant plant growth, cause seasonal variation in decomposer activity in soils. Soil organisms have the advantage of a **buffered** climatic environment. Nevertheless, they are often considerably controlled by the soil climate either in terms of which species can live under the prevailing conditions, or the rate of action of these decomposers, or both. Soil organisms are especially influenced by soil temperature conditions, and by the availability of water in the soil (either too little or too much). The latter has the effect of restricting the availability of air in the soil. Air supplies the oxygen essential for aerobic processes – respiration – which is the way in which most heterotrophs release chemical energy from their food. Aerobic respiration is not only the most common mechanism for liberation of energy, but also the most efficient. Other mechanisms which release chemical energy from organic debris use ions such as ferric iron or nitrate which have a lower redox potential (see Box 4.2) and thus release less energy. Furthermore, these anaerobic processes do not result in a complete breakdown of organic residues.

Box 4.2

Redox and redox potential

Redox, or more fully the oxidation-reduction potential of a compound (E_h, measured in mV), is an indicator of the energy level of that substance. Strictly, redox measures the ability of a substance to reduce other compounds, or to be oxidised. We can consider reduction as the gain of an electron, and oxidation the loss of an electron in a chemical reaction as a balancing pair of reactions. The acronym OILRIG is useful (oxidation is loss, reduction is gain).

The most important reactions for living organisms are photosynthesis, which is the reduction of carbon, and respiration, which is the oxidation of carbon-based compounds. However, other substances that are common in ecosystems are also good reducers and oxidisers. Ammonium (NH_3) is a strong reducing agent, while nitrate (NO_3^-) has a high oxidising capability. Thus the importance of redox is related to the energy transformations that occur as a result of oxidation and reduction in living systems and their physical environment. Life depends upon the ability of organisms to store energy chemically and to release that stored energy as and when it is required by the organism's metabolism.

Redox potential is a useful overall measurement that can be made to assess the overall abiotic ecological environment of a location. It is easily measured by using a meter that shows the concentration of electrons in the location. The lower the concentration of electrons, the more positive the location, which thus has a higher oxidation potential. Reducing conditions will be shown by negative potential values. An example of an ecosystem in which E_h can be sharply reduced is the flooded soils of a marsh habitat. In flooded soils most of the soil pores are filled with water, reducing the rates of gas diffusion through the soil approximately by a factor of four. As a result it is difficult for oxygen to move through the soil, and often beneath a surface oxidized zone just a few millimetres in depth, the soil; is anaerobic. In such conditions a succession of anaerobic reduction processes go on, each in turn further reducing the redox potential as a result of the reduced products formed (Laanbroek 1990; Ricklefs 1990):

E_h (mV)
- Disappearance of oxygen   +330
- Disappearance of nitrate   +220
- Appearance of managanous ions   +200
- Appearance of ferrous ions   +120
- Disappearance of sulphate   −150
- Appearance of methane   −250

The net result is that the soil becomes an increasingly hostile environment for plant root survival as reduction proceeds. The appearance of toxic substances such as methane, coupled with the absence of oxygen makes root survival increasingly difficult in heavily reduced soils. As a result, wetland plants which colonise such high-stress habitats have had to evolve special adaptations (such as aerenchymatous root tissues, which pipe air down to the roots from the foliage) to cope with these conditions.

pH is a powerful control upon decomposition of organic matter, acting in two ways. First, the strong reducing conditions associated with acidity trigger reactions such as **chelation**, which greatly reduce nutrient availability. Second, in strongly alkaline conditions, although some nutrient ions may be abundant, others may be scarce, and the ions are often located in precipitated salts unavailable to plants. In both extremes soil biotic activity is inhibited. For example, many cellulose-decomposing bacteria are unable to tolerate pH values below about 5.0, and earthworm activity is much reduced below about pH 4.5. In arid environments, in which soil pH may exceed 9.0, all biological action is severely reduced, and organic debris is scarcely decomposed at all. Thus ideal soil conditions for the most efficient decomposition of organic detritus are generally in the range pH 5.5–8.0, with the optimum being about neutral (7.0). Soil pH is related to several factors, including the nature of the action of decomposers, but is primarily controlled by three external, forcing factors. These are the nature of the organic material input, the nature of the input of weathered parent material, and climatic conditions, particularly the characteristics of water movements in the soil.

One of the best indicators of the way in which organic debris will break down is the ratio of carbon to nitrogen (C/N) in the material. Vegetation which produces debris with a high C/N, in the range 100/1, is derived from cellulose-rich products and is resistant to change. Material with a low C/N, in the range 10/1, is much more susceptible to rapid and complete decomposition. The nature of the living vegetation that contributes this debris is itself controlled by a whole complex of environmental factors which are in turn interrelated. Climate as well as soil conditions shape a cyclical system of interactions between plants and soil, which lie at the heart of the nutrient cycling system. The nature of parent material affects nutrient cycling in two ways. First, it has an influence on the soil pH regime through the chemistry of the mineral fraction of the soil. Second, the particle size of the mineral fraction, **soil texture**, has a significant impact upon the transmission and retention of water in the soil profile. Thus climate alone is not the determinant of soil water conditions. The water budget is the balance between upward movements of water driven by evapotranspiration, and downward movements driven by drainage under gravity. These in turn are related to inputs of water through precipitation inputs, and loss of water through evaporation, influenced by temperature, atmospheric humidity, wind-speed and insolation. Actual soil water and therefore soil air conditions are also influenced by soil texture, variations in which influence the amount of water which can be held temporarily in the soil and the rate at which water drains through the soil. The net movement of water through the soil affects soil chemistry, and thus soil biology and nutrient cycling, by removing soluble salts from the soil by the process of leaching, or following evaporation, concentrating salts in the upper soil horizons.

Cycling of nutrient supply to plants is absolutely crucial for ecosystem functioning. Inevitably, there are long-term losses of nutrients from all types of soil. Although this may be accelerated by human actions, some loss is normal and natural. Under natural conditions this does not generally lead to nutrient depletion, since as well as losses there are inputs from outside the nutrient cycling part of the ecosystem. Although it could be argued that these slow changes are a part of the cycle, because the time scale is so different from the core of cycles it is helpful to consider these separately. There are two long-term pools for material lost from the cycle system – the world ocean, ultimately much of which is converted to sediments, or the atmosphere. New inputs are derived from these external pools, for example, by rock weathering in the case of phosphorus or biological fixation in the case of nitrogen. Both of these are considered in more detail later in this chapter. For nutrient cycling to work effectively there has to be some sort of nutrient storage system in the soil which prevents soluble material from being taken away from plant roots by soil water movements, while at the same time allowing plants to use the nutrients. Such a mechanism exists, and is based on the colloidal properties of soil. Although the colloidal store is a small pool, its significance to nutrient cycles is considerable (see Box 4.3).

Box 4.3

Colloids and the soil

What are colloids?

Colloids are physical mixtures of materials, in which the individual particles are so small that the mixture is in a stable condition, even though it is not a chemical compound. The natural world is particularly rich in colloids. Examples include milk, blood, clouds and the soil. Colloids are made up of material which are in different physical states or phases. Any combination is possible. Colloids which have predominantly liquid-like properties are in the gel state, while those which appear more solid are referred to as gels. The intimate mixture of substances which constitute a colloid not only give a persistent condition, but also give colloids particular properties. These are related to the tiny size of the particles which gives them distinctive electro-chemical properties, and are the basis of their physical persistence.

Colloids in the soil

Soil colloids are drawn from two main sources. Tiny mineral particles, in the *clay fraction*, as shown in the distribution of size classes below, are the first type.

Fraction or size class name	Particle diameter (mm)
Sand	2.00–0.05
Silt	0.05–0.002
Clay	< 0.002

Some clay particles are tiny fragments of rocks and their constituent minerals. However, the majority are alteration products following rock decomposition. These are called *clay minerals*. There is thus a distinct difference in definition between the clay fraction, a particle size, and clay minerals, a particular chemical product. In many cases the majority of clay fraction particles are clay minerals.

The second major class of soil colloids is humic material. *Humus* is also an alteration product, formed by the breakdown of organic detritus, and re-synthesis of some of these breakdown products with organic residues. There are different types of humus and different types of clay minerals, which have significantly different colloidal properties.

Importance of soil colloids

Soil colloids are important because these can act as a temporary store for available nutrients in their simple ionic, soluble form. This is the property of *adsorption*. Adsorption is the attraction of positively charged ions or cations, to the surfaces of soil colloids. Soil colloids have weak negative electro-static charges on their surfaces. The total capability of soil colloids to attract and hold cations is termed its *cation exchange capacity* (CEC). The adsorption process resists loss by leaching of soluble nutrients, while still making nutrients available to plants which are able to obtain adsorbed nutrients via the soil solution. Not all nutrients are in the form of cations. For example nitrogen and phosphorus are largely used by plants in anionic form (nitrate $[NO_3^-]$, phosphate $[PO_4^{3-}]$). The soil colloid store, though small in absolute size compared with other pools in the nutrient cycle, is of considerable importance to ecosystem productivity, and to the human resource value of soil.

Material cycling

A hydrosphere-based cycle: the hydrological cycle

The basic hydrological cycle system is shown in Figure 4.3. The material of the cycle is water, in all three physical states – ice, liquid water and water vapour. The amount of energy involved in the changing state of water is very large indeed. Energy used in changes of state plays an important role in the global circulation of energy, which drives the Earth's climate. The energy source which drives the hydrological cycle is solar radiation. Over the temperature range found across much of the Earth's surface, water can exist freely in liquid and vapour states. As water is relatively plentiful in the biosphere, the changes of state which characterise the hydrological cycle function might seem to have more significance for climatic and geomorphic systems than for ecosystems. But this is not so. Apart from the fact that climate and land forms are a major element in the abiotic environment of ecosystems, water in terrestrial environments is often relatively scarce. Without a rapid and effective cycling system most parts of the land surface of the planet would be unable to support any autotrophic plants, and thus would be devoid of life. Therefore the ambient temperature of the Earth's surface is critical, by allowing

Figure 4.3 Basic hydrological cycle

water to be moved quickly from the world ocean to land surfaces through the processes of evaporation and condensation.

Water in the world ocean contains about 3.5 per cent of dissolved salts, mainly sodium chloride. Although marine plants are fully adapted to use this type of water, terrestrial plants require and receive much purer water. The process of evaporation affects only pure water, leaving behind soluble salts. Rain-water picks up some dissolved material and solid particulate aerosol material, such as dust, in its passage through the atmosphere. However, compared with sea water it is relatively chemically pure. This is the condition required by nearly all land plants. Environments which have water of varying chemical quality over short time scales, such as estuaries, are particularly stressful for plants. This does not mean, however, that such ecosystems are unproductive, because autotrophs which have evolved specialised survival strategies to cope with salt stress (e.g. salt marsh plants) can flourish there, and ecosystem productivity can be quite high (see Chapter 5). However, for many land plants a major environmental constraint is securing a water supply which is relatively chemically pure, and is sufficient in quantity to sustain plant growth.

Pools in the hydrological cycle are of very different sizes. The soil water and atmospheric pools in particular are very small indeed, in relative terms. However, although they total less than 0.01 per cent of all water in the hydrological cycle, this is a large absolute quantity. Further, the exchanges between the surface of the Earth, both land and sea, are rapid, so that residence time in these pools is short. An essential concomitant of unequal pool size is different rates of transfer between the pools, and different residence times. This is necessary to maintain long-term continued functioning of the hydrological cycle. Thus it has been estimated that all the world's water is involved in photosynthesis and respiration about once every two million years (Cloud and Gibor 1970). Oxygen is recycled every 2,000 years and the residence time of CO_2, the least abundant major nutrient, has an atmospheric residence time of about 300 years (Cloud and Gibor 1970). Residence times in the lithosphere for macro-nutrients are measured on geological time scales.

The soil, ground and surface water pools are essential for life. They act as temporary stores which provide direct and indirect links to the plant. These pools are sustained by rainfall and other forms of precipitation. Most evaporation and rainfall occurs as a simple loop from the ocean to atmosphere and back to the ocean again. However, atmospheric circulation causes some rainfall over land. This balances overland runoff, which returns on land to the sea, and evapotranspiration which returns soil water and transpired water directly to the atmosphere. The transpiration path in land areas where water is relatively abundant is the major means whereby water is returned directly to the atmosphere; transpiration is also the main way in which plants dissipate heat energy. Surface water bodies in terrestrial environments, such as rivers and lakes, account for a very small part of the water in the hydrological cycle. Biological productivity in many of these, but by no means all, is low, as these water bodies are often nutrient deficient. However, where nutrients are available, in ecosystems such as eutrophic wetlands or lakes, productivity may be high.

Ice-caps may at first sight appear to have little effect on ecosystem functioning because they are largely devoid of plant life (but see Chapter 5 for a description of some of the stress-tolerant ecosystems associated with ice). However, their sheer size gives them importance. Representing about 2 per cent of the total water in the hydrological cycle, even quite small changes in their size will affect the volume of the world ocean, and thus sea-level. Ice-caps also play an important role in controlling climatic patterns. There is thus an interactive relationship between ice-caps and climate. This issue exemplifies one of the most significant current environmental issues, that of human-induced environmental change, and the consequences of this for ecosystem function. Water pollution has already been mentioned, and at local or regional scales this is a major environmental problem which affects human well-being and ecosystem function. At a global scale, human-induced climatic change, so-called global warming, caused by a human 'short-circuiting' of the carbon cycle (discussed in Chapter 11), has the most profound implications for ecosystems. Climatic change associated with global warming will not only change the thermal environment, but will also modify spatial and temporal precipitation patterns. By its effect on plant life, this disruption of the functioning of the hydrological cycle will have a major impact upon natural ecosystems, and upon agriculture. Following the report of the 1996 International Scientific Committee on Climate Change the existence of global warming is now a scientifically accepted reality. Since then a mass of scientific evidence has been presented to support the reality of humanly induced global climate change, even if there is still debate on its amount and rate (Drake 2000, 137–202; Beeby and Brennan 2004, 236–250). At present it is not possible to predict accurately its regional dimensions, so that future consequences for vegetation remain unclear.

A biosphere-based cycle: the carbon cycle

The cycling of carbon is closely linked to energy flow through ecosystems. Indeed, we sometimes refer to life on Earth as being carbon based, because the organisation of energy upon which life depends is done, to a large extent, through the combination and breakdown of carbon compounds. Figure 4.4 outlines the carbon cycle. The carbon cycle

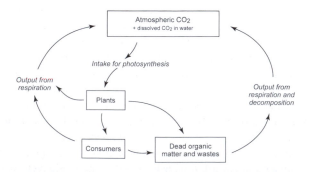

Figure 4.4 The carbon cycle: the fundamental cycle

has four types of pools; like the hydrological cycle, these are of very different sizes. To understand the operation of the carbon cycle properly, we need not only to examine pools and links within the biosphere but also to include carbon located in the near-surface lithosphere. Through geological processes such as lithogenesis and weathering, this C links with other pools which fall within the biosphere's boundaries. The atmospheric pool links directly to the oceanic and biological pools, and flux between the atmospheric pool and these pools is rapid. Linkage with the geological pool is indirect and flux slower, though human actions over the past two centuries have modified this, with increasing implications for all biospheric processes. The atmospheric pool comprises less than 0.5 per cent of the total amount of carbon in or close to the biosphere and its environmental systems. Nevertheless, the speed and rapidity of transfers between the atmospheric pool and the biosphere are such that shortage of carbon dioxide is rarely a limiting factor on primary production. The lowest part of the atmosphere, the troposphere, is a fairly constant mixture of gases, of which carbon dioxide comprises 0.38 per cent (estimates give a value of about 0.28 per cent before large-scale industrial activity and 0.38 in 2004: data based on Houghton *et al.* 1996). However, this is quite sufficient to sustain terrestrial plant productivity, and other limiting factors, such as water or nutrient supply, normally act as controls on production rates. The geological time scale link between atmospheric composition was discussed in Chapter 1. The current cycle pattern is the outcome of very long-term change and adjustments, both within the biosphere and in crustal areas close to the life zone.

The supply of carbon is not a problem for most aquatic primary producers. Carbon dioxide is soluble in water to the extent that again, limiting factors other than carbon supply, principally nutrient availability, generally act as controlling factors on primary production. Indeed, the world ocean contains much more carbon dioxide than does the atmosphere, so that generally in aquatic environments CO_2 is plentiful. In a few aquatic environments in which mixing of air with surface water by wave action is limited, the supply of dissolved carbon dioxide may be limited. However, such habitats are restricted in size. In low pH waters, such as acidified lakes, the supply of dissolved carbon dioxide may be limiting to growth, and plants living submerged in such systems have developed specialist adaptations to get at extra supplies of CO_2, for example, from the sediments (see Chapter 2). Deep water may also lower carbon dioxide availability, but light extinction, which occurs quite rapidly with increasing depth, means that photosynthesis cannot take place in such environments. Oceanic waters contain a considerable proportion of all carbon in the biosphere and the near-surface lithosphere, and chemical/biological exchanges between the ocean and the atmosphere account for about 60 per cent of the global total of biosphere–atmosphere interactions. Most of the carbon in the world ocean is in the form of carbonates. Some of this is converted into carbon dioxide, and respiration by marine organisms contributes carbon dioxide to the atmosphere too. Marine photosynthesis uses dissolved carbon dioxide obtained directly from water, so that the gas form is the available state for both land and aquatic autotrophs. A large part of the oceanic carbonate is converted into sediment over geological time scales. Such sediments include limestone and chalk, which through the long-term evolution of the Earth's crust may be raised and weathered, thereby releasing carbon dioxide into the atmosphere. This exchange is very slow, and is insignificant compared with biological cycling paths.

Very large quantities of carbon are locked in carbohydrate-rich organic material which has been converted by geological processes into sediments or trapped fluid residues. These are the fossil fuels. The global pool of coal, oil and natural gas accounts for more than 20 per cent of the biosphere and near lithosphere total of carbon. Under normal circumstances release of this material, which is located in the upper part of the Earth's crust beneath both land and sea surfaces, is very slow. However, human use of these materials has accelerated at such a rate over the past 200 years that the almost instantaneous return of carbon dioxide to the atmosphere by combustion is having a significant effect on the carbon cycle, and thus upon environmental conditions in general. Burning fossil fuels releases the energy which had been fixed by plants in bygone geological times. It also decomposes geologically altered carbohydrates into carbon dioxide, water and some residual compounds. The resultant increase in atmospheric carbon dioxide causes increasing temperature in the troposphere due to the infra-red energy capture properties of CO_2. As levels of carbon dioxide can be measured with some precision, we know that atmospheric content has increased about 15 per cent in the past 200 years. The actual climatic outcome of this is not yet clear, but it is a major environmental concern. These issues are discussed further in Chapters 10 and 11.

The final pool in the carbon cycle is the carbon content in current organic matter. This includes both currently living plant and animal tissue, though the structure of ecosystems means that the vast majority of biological material is in, or derived from, autotrophs. The amount of organic matter per unit area varies considerably over the surface of the Earth. As has been explained above, CO_2 is rarely the principal control on primary production. Thus it is other constraints on photosynthetic activity, such as heat conditions, availability of water or of nutrients, which act as limiting factors. However, these constraints have little effect on the critical paths for biological activity in the global cycling of carbon. Because the critical paths are between the atmosphere and primary production (and thus all trophic levels in ecosystems) via photosynthesis and respiration, there is an effective overall balance in this part of the system. Disturbance to other links can disrupt the system. Furthermore, the particular roles of certain specialised biota in ecosystems are vital to the maintenance of the carbon cycle. The roles of specialised decomposers were noted earlier in this chapter. Without the efficient action of cellulose decomposers such as bacteria, fungi, ants and termites, much carbon would remain locked in unavailable form in wood. It is interesting to note that different types of wood digesters predominate in different climatic environments. The decomposers act at different rates so that in hot, wet, tropical ecosystems little organic residue, however woody, will persist for more than a few months, whereas in colder or drier conditions which inhibit the action of decomposers, woody material may remain relatively unaltered in substantial accumulations of plant litter in the soil for many years. Similarly, organic sediments in aquatic environments lacking in oxygen would become another carbon sink were it not for the activities of anaerobic bacteria. These organisms can oxidise organic detritus, typically with sulphur in an oxygen-free environment. Although less efficient than aerobic action, in the sense that breakdown of organic material is slower and incomplete, anaerobic decomposition is important in returning CO_2 to the atmospheric or aquatic pools, where the gas can be recycled in photosynthesis. Although important to the overall functioning of the carbon cycle, quantitatively these paths are much less important than the biological links which depend upon photosynthesis, or carbon fixation, and its breakdown by respiration.

An atmosphere-based cycle: the nitrogen cycle

Plants and animals require nitrogen as components of nucleic acids and proteins. In absolute quantitative terms, nitrogen is the macro-nutrient which is required in largest amounts. The nitrogen cycle is shown in outline in Figure 4.5. One of the most striking features is the huge pool of nitrogen in the atmosphere, which is 78 per cent composed of this gas. However, atmospheric nitrogen (N_2) is not only in a form which is unavailable to most autotrophs, but is also a very stable molecule that requires significant amounts of

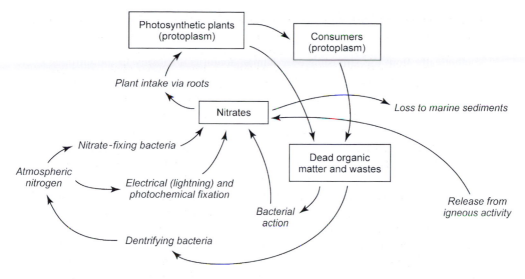

Figure 4.5 The nitrogen cycle: an atmospheric link cycle

energy to convert into forms which can be used by plants. A relatively small amount of atmospheric nitrogen is fixed, or converted into soluble nitrate by lightning discharges. Biological links between the giant atmospheric pool, soil and ocean are more important.

Biological fixation of nitrogen is carried out by a small number of species of bacteria. Many of these are free-living (for example, the nitrogen-fixing cyanobacteria such as *Anabaena*, in the phytoplankton of lakes: see Chapter 7). Others form symbiotic associations with plants. The best known are *Rhizobium*, *Frankia* and *Azotobacter*, which live in colonial groups in the roots of legumes and a number of other genera of vascular plants (e.g. *Frankia* is a symbiont on the roots of alder trees, *Alnus*). The resultant nodules are the sites of much biological nitrogen fixation, and represent one of the most important symbiotic relationships which exists on Earth. Without bacterial N-fixation, primary production in water and on land would be severely limited. Lack of nitrogen is a common limiting factor in many ecosystems in both realms. Some nitrogen is lost from the system to deep ocean sediments, though some is brought in from volcanic activity. Loss of available nitrogen from land areas to the ocean is inevitable, since most available nitrogen in nitrate form is both soluble and anionic, and thus not held by soil colloids. Much of this nitrogen is used by marine autotrophs, since nitrogen is particularly scarce in sea water. Some of this nitrogen is moved back to the land by sea birds.

The main form in which nitrogen is used by autotrophs is as the anion nitrate (NO_3^-), though a significant amount is in cationic form as ammonium (NH_4^+). Organic debris, particularly non-woody material and animal residues, often contains significant amounts of nitrogen. This is broken down by denitrifying organisms, chiefly bacteria, in the soil and ocean and returned to the atmospheric pool. As with the fixation path, considerable amounts of energy are liberated in the stepwise breakdown of complex nitrogen-rich organic residues through denitrification. The nitrogen cycle exhibits a high degree of stability, via the biological links to the huge but almost inert atmospheric pool. The system is stabilised through feedback provided by the microbiological nitrifying and denitrifying paths. However, stability is threatened, as we have noted in other cycles, by human actions. As nitrogen is frequently a limiting factor, intensive farming commonly uses large amounts of nitrogenous fertiliser. This is fixed synthetically from the atmospheric pool using large amounts of electrical energy. Production of nitrogen fertiliser, in the form of a suitable nitrate salt, is a major sector of the global agricultural chemical industry, and economies of scale, together with the fact that the cost of the input electricity is less than the value of the increased output which results from the use of fertiliser, has meant that very large amounts are used. This use of nitrogen derived synthetically from the atmosphere by human action is a subsidy to the nitrogen cycle system.

The problem which may result from the large-scale use of nitrogenous fertiliser lies less in the energy used in the manufacture, but more in the fate of the nitrogen applied to the soil. As previously noted, nitrates are anions and are thus not adsorbed by soil colloids. Nitrates are highly soluble, and thus easily leached from the soil by water draining through the soil profile. Excess nitrates thus quickly accumulate in surface water bodies, such as rivers and lakes, and in sub-surface groundwater. These losses, together with nitrogen which is removed in crops, the fate of which is complex, are known as drains. Subsidies and drains not only upset the balance of the nitrogen cycle but also cause environmental problems. These issues are discussed more fully in Chapter 8.

A lithosphere-based cycle: the phosphorus cycle

The phosphorus cycle has both similarities and differences when compared with the nitrogen cycle. Figure 4.6 shows the system of pools and links involved in the phosphorus cycle. Like nitrogen, phosphorus is a macro-nutrient, but is required in rather smaller quantities. Phosphorus has a wide variety of biological functions including roles in nucleic acids, cell membranes and skeletal systems. Phosphorus plays a central role in the fundamental energy transfer processes of photosynthesis and respiration in cells, via molecules such as adenosine triphosphate (ATP). Phosphorus is relatively scarce in

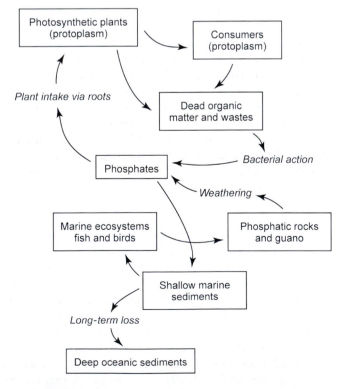

Figure 4.6 The phosphorus cycle: a solution cycle

most environments, and many organisms have mechanisms whereby phosphorus may be stored internally. Phosphorus deficiency is often a limiting factor in terrestrial environments, and is the most common limiting factor in aquatic, particularly freshwater environments. Apart from dust aerosols and sea-salt spray there is no atmospheric phosphorus, so that the cycle is based on interaction between the biological components of ecosystems, soil and water. The cycle is therefore simpler and less controlled by biological feedback loops than the nitrogen cycle. Organic detritus is converted by decomposing bacteria into phosphate (the available form in solution in the soil water as PO_4^{3-}), which is then reutilised by autotrophic plants, and the higher trophic levels in ecosystems. This is the main loop within the cycle. This organic loop operates at a much more rapid rate than the links with the oceans and geological substrate. It is, however, limited in size in relationship to global biological demand for phosphorus. The crucial organic loop is thus more fragile and less self-adjusting than the main mechanisms in the nitrogen cycle.

The availability of phosphorus is highly sensitive to pH of the substrate. In both acid and alkaline conditions, phosphorus is converted to insoluble or very weakly soluble compounds which are unavailable to plants. Perversely, although phosphorus in its available form (phosphate: PO_4^{3-}) is anionic and thus not adsorbed by clay, as pH departs from neutrality, phosphates are bound chemically with clay particles, and become unavailable to plants in many soils. The active pool of phosphorus, cycling between living and dead organic material and soil and aquatic pools of available phosphate, is limited, and tends over time to be depleted by loss of phosphate salts to deep sediments beneath the ocean. These sediments, over the tens of millions of years of geological time periods, are converted into sedimentary rocks, and may be exhumed from the oceans by mountain-building processes. As these mountains are attacked by geomorphic processes, phosphates are made available to autotrophs by rock weathering. The time scale of this loop is very long. A locally significant and more rapid loop in the cycle is the transport of phosphate from the oceanic sink, by the consumption of marine organisms by birds, the guano of which is deposited on land. These deposits have been exploited by humans seeking sources of phosphate-rich material to use as fertilisers. Modern phosphate fertilisers are derived from processed phosphorus-rich rocks. It is notable that the rate of human use of this source of nutrients, built up as part of the natural cycling system, is much faster than the rate of replenishment. Other methods of boosting phosphate availability, such as using fishmeal-based fertiliser or mining and processing phosphate-rich rocks, also lead to damage to ecosystems, nutrient-cycling systems and the environment.

Human actions to boost the supply of available phosphorus may have serious environmental impacts. Mining of phosphate-rich rock itself may be locally damaging. However, it is the delivery of excess phosphate into natural water bodies which causes more serious problems. This may occur as a consequence of leaching of phosphatic fertilisers from the soil into drainage water. As is analysed more fully in Chapter 8, eutrophication may result from natural balances being upset by the entry of phosphate into water bodies. Leached fertiliser is not the only cause of eutrophication. Discharge of untreated or partly treated effluent, which is particularly rich in phosphate, is a common cause of eutrophication, which by the disturbance of normal patterns of primary production in water bodies, can have catastrophic consequences on both aquatic biology and water quality. Recently there has been concern about the loss of phosphates, and other nutrients, by accelerated soil erosion. Research has shown that the rates of phosphorus loss from soils in the USA and Australia as a result of erosion and leaching are much greater than was previously believed (Cutter and Renwick 1999).

The nitrogen and phosphorus cycles are examples of the two general types of macronutrient cycles (Deevey 1970). Cycles such as the nitrogen cycle which have an atmospheric link tend to cycle nutrients more quickly than do those which are effectively confined to ocean and land. Deevey (1970) called these carboxylation and soluble element types respectively. Although there is no simple relationship between the absolute size of various pools in different nutrient cycles, the rates of cycling and the *relative* availability of nutrients, it is notable that the nutrients, which are required in the greatest absolute amounts by ecosystems, involve atmospheric transactions. Not only does the atmospheric pool provide a large, easily accessible source of these nutrients, but also the biological organisms, which provide the links with the atmosphere, are responsive feedback-controlled mechanisms which provide stability to the rapidly flowing cycle of nutrients. Soluble element cycles are much more readily disturbed by human as well as natural circumstances, and macro-nutrients in this category are often limiting factors in particular geographical environments. The implications of this for global vegetation patterns and ecosystem function are discussed briefly in the next section.

Material availability as a limiting factor in ecosystem functioning

Availability of nutrients is a major control on global patterns of primary production, and on the nature and functioning of ecosystems, both at a global scale and at smaller spatial scales. This chapter has shown that the continued supply of essential nutrients, upon which life depends, requires the uninterrupted functioning of nutrient cycles. Nutrient cycles include biological paths which assemble and break down the nutrient building blocks of living material. In cycles of those nutrients which are required in large absolute amounts by autotrophic plants, biological cycling routes are important. Through acting as feedback mechanisms, primary producers and decomposers regulate and stabilise the cycles. Human activities which modify nutrient cycling through overuse of nutrients constitute an environmental threat through overloading nutrient-cycling systems. Intensive agricultural systems are heavily subsidised by use of synthetic fertilisers. This applies particularly to nitrogen, phosphorus and potassium. It should be remembered that supplies of nutrients can be increased by use of natural fertilisers, such as composted organic matter, or animal excreta. However, synthetically produced fertiliser is relatively cheap, and is easy to handle. Reliance on synthetics together with continuous monoculture of crops, both of which are common in intensive agriculture, leads to depletion of soil organic matter. When taken with the excessive application of synthetic fertilisers which are water soluble, the inevitable results are large-scale leaching losses and water eutrophication problems.

One of the ironies of this environmental problem is that, by one measure, intensive agricultural systems are less efficient than less intensive methods. The amount of energy captured through cropping declines per unit of fertiliser applied in heavily subsidised systems. This means that, although total production increases, there are diminishing additional returns in relation to fertiliser use. This issue is discussed in detail by Pimental and Pimental (1979). That this happens relates to the economic value of the output from the system. These are greater than the cost of the inputs. A further complication is that the economic value of inputs and outputs changes over time. In part this relates to the scarcity of each, but it is also affected by the complexity of modern economic production systems, and is hard to project in the medium term. It is equally difficult to make quantitative measures of ecological values. The underlying problems are the different time scales involved in human economic systems and in natural ecological systems, and the pressure imposed on the Earth's resource base by contemporary human society. For many people in different parts of the world, increasing food production has a much higher immediate priority than sustaining environmental systems. Some aspects of these and related problems are examined in Chapters 10 and 11.

Summary

- This chapter explains what is meant by nutrients, and how nutrients are made available for plants.
- The role of soil in the cycling of nutrients for plants in terrestrial environment is explained.
- Nutrient cycling, which is an integral and vital part of ecosystem functioning, is explained in general terms, and through specific consideration of the cycling systems for water, carbon, nitrogen and phosphorus.
- General types of the macro-nutrient cycling system are identified.

Discussion questions

1 For what purposes do autotrophic plants require nitrogen? What happens to plants which suffer from a deficiency in the supply of nitrogen?
2 Find summaries that include diagrams of the sulphur and potassium nutrient cycles. What types of nutrient cycle are these?
3 What nutrient subsidies and drains occur in livestock farming? Are there any significant differences in subsidies and drains between intensive animal rearing and extensive ranching? Are there differences between pastoral systems in more and less economically developed countries?

Further Reading

See also

Stressed ecosystems, Chapter 5
Disturbed ecosystems, Chapter 6
Human Impacts on ecosystems – impacts on trophic structure, Chapter 9
Large-scale impacts on ecosystems – the increasing effects of humans, Chapter 10
Global environmental change and consequences for ecosystems, Chapter 11
The movement and availability of nutrients are so central to the functioning of ecosystems, and to functional ecology, that this is one of the most important elements of the ecosystem concept. Most parts of this book, and many issues in environmental and ecological management, are related to material cycles in the biosphere.

Further reading in the Routledge Environmental series

Environmental Biology
Natural Environmental Change

General further reading

The Biosphere. I. Bradbury. 1991. Belhaven, London.

Chapter 2, 'The chemical basis for life', gives a useful overview of the chemical properties of the biosphere, written for the non-specialist chemist.

The Biosphere. D. Flanagan (ed.). 1970. Freeman, San Francisco, CA.

This is a reprint of the issue of the journal *Scientific American*, September 1970. It has excellent essays on all aspects of nutrient cycling, written by leading specialists in the field.

Ecological Principles and Environmental Issues. P.J. Jarvis. 2000. Prentice Hall, Harlow. Chapter 2, 'Food, energy and nutrients', provides a useful overview of both energy and nutrient cycling and develops issues raised in the specific context of agro-ecosystems, and environmental management in general.

5 Ecosystems in high-stress environments: meeting environmental challenges

Stressed ecosystems challenge the survival of the organisms which occupy them by imposing extreme heat or cold, dryness, or lack of light or nutrient (to take some of the major causes of stress). In response, stress-tolerant organisms must invest in a range of expensive-to-build adaptations to permit survival. The very existence of these adaptations restricts them to the stressed habitat. This chapter explains why and how such restricted distributions occur, and looks at some of the ways in which plant and animal species have adapted themselves to life in ecosystems experiencing such harsh survival conditions. It also describes how stressed ecosystems may be naturally resistant to bio-invasion and the problems this causes for biodiversity maintenenance. This chapter covers:

- **Defining and measuring environmental stress**
- **Effects of stress on animal populations in stressed ecosystems**
- **Strategies for adaptation in stressed ecosystems**
- **Stress tolerance strategies in plants**
- **Role of environmental stress in countering bio-invasions**

Defining and measuring environmental stress

In Chapter 2 we defined environmental stress as any factor which tends to reduce the efficiency of functioning of one or more key physiological processes in the organisms occupying a given ecosystem. The organisms which occupy high-stress ecosystems must have the right combination of adaptive characteristics to meet the challenges which the environment offers to survival there. In the case of green plants, which rely absolutely on photosynthesis to produce the food they need to survive, anything which reduces the efficiency or rate of photosynthate accumulation is a source of stress. Ecosystems which impose this sort of stress on plants include:

- the understorey habitat of forest ecosystems (where shade is the main source of stress)
- high mountain ecosystems (where cold and high winds combine to stress the plants by coupling low metabolic activity with so-called 'physiological drought' produced by the desiccating effect of the winds)
- salty ecosystems, such as salt marshes and salinised irrigated farmland (where, again, accumulation of too much salt in the cells of the plants causes them to dry out through osmotic processes)
- hot desert ecosystems (where direct drought stress is the biggest problem for plant survival)
- nutrient-stressed systems, such as oligotrophic lakes or areas of land with low-fertile soils, where plant growth nutrients are in limited supply.

Once we have identified the key sources of stress affecting plant growth in a given ecosystem, quantifying the intensity of stress is simply a matter of physical measurement.

For example, the intensity and seasonal duration of heavy shade beneath a deciduous forest canopy can be recorded by installing a data logger on the forest floor, linked to a pair (above and below the canopy) of light meters sensitive to photosynthetically active radiation (PAR: the wavelengths of light in the range $0.38-0.78$ µm, within which chlorophyll shows major absorption peaks in the blue and red bands). These wavelengths of light are selectively absorbed by the tree leaves in the overhead canopy. Beneath a dense oak forest canopy the PAR intensity at ground level may be less than 1 per cent of the incoming energy above the forest canopy.

As a second example, in a North European salt marsh ecosystem the amount of salt present in the soil is a direct function of two things (Crawford 1989). One of these is the probability of inundation by sea water, which is linked to the tidal cycle, and diminishes as we move higher up the marsh away from the sea. At the top end of the marsh only the very highest of high spring tides will cover the soil, probably only a few times each year. The diurnal inundation by the tide may also be by water of varying salinity if the salt marsh is located in an estuary, depending on the height of the tide and the amount of flow in the river: a neap tide plus a flood in the river may combine to give only brackish or even freshwater covering the marsh at high tide. But if there is a large spring tide, coupled with low flow of freshwater from the river (e.g. during a summer drought period) then the marsh may be covered with near full-strength sea water. Thus the intensity of salt stress which the plants face may vary substantially. The second important factor is the exposure of the marsh to wind and wave action. Marshes located in more exposed areas (such as a promontory) are exposed to higher winds and bigger waves, which both carry the sea water further inland and so higher up the marsh, and also drive salt-laden sea spray further inland.

The net result of these factors is that a strong gradient of saline conditions occurs in the soil of a salt marsh. Lower down the shore salt concentrations in the soil (which can be directly measured at low tide using a salinometer) are similar to the full salt concentration of the adjoining sea (around the British Isles averaging thirty-four parts per thousand, lower in estuarine conditions). Higher up the marsh the salt concentration progressively declines, and so does the intensity of salt stress experienced by the **halophytic** ('salt-tolerant') plants occupying the marsh habitat. A complicating factor is that waterlogged salt marsh soils low on the salt marsh gradient (i.e. those flooded more regularly and for longer by the tide) usually have strongly reduced conditions. This can be measured directly by using a redox probe (see Chapter 4) to determine the likelihood of oxidising conditions in the soil. Low redox values indicate reduced conditions often highly deficient in oxygen (**anaerobic**). Such reduced soil conditions are highly toxic to the survival of the roots of a number of halophytic upper-marsh plants which would otherwise be perfectly capable of surviving lower on the marsh if salt stress was the only stress influencing the ecosystem. The net result of the combination of stress conditions influencing plants in a salt marsh ecosystem is to produce a strong spatial pattern in the vegetation. Specialist stress-tolerant species achieve near complete domination in those areas of the marsh to which they are adapted, but are completely excluded from other areas by more competitive, non-halophytic species. There is a very clear zonation of vegetation in a typical north German salt marsh (Box 5.1), with plants strongly tolerant of salt stress and root anaerobiosis occurring low on the marsh (e.g. *Spartina* grasses), while those occupying the top end of the ecosystem have much lesser tolerance of salt marsh stress conditions. Box 5.2 describes some of the physiological mechanisms used by halophytic plants to resist the effects of salt stress.

Box 5.1

Salt marsh zonation

Found in temperate estuaries and soft-sediment marine shores. Tidal levels determine three-zone marsh system:

- Accretion zone (mean low water MLW – mean sea-level MSL): soft sediments accumulated around seaweeds, e.g. bladderweed (*Enteromorpha*) and debris; level of marsh surface starts to build up
- Stabilisation zone (MSL – mean high water MHW): pioneer species invade, e.g. glasswort (*Salicornia*), cord grass (*Spartina*)
- Emergent marsh (MHW – extreme high water of spring tides EHWS): plant litter plus sediments build up marsh until

either: *Spartina* dominates, forming large monospecific stands of vegetation

or: mixed marsh develops, supporting a more diverse plant community, dominated by plants such as sea arrow-grass (*Triglochin*), sea lavender (*Limonium*) or salt marsh grass (*Puccinellia*)

Box 5.2

Strategies for surviving salt stress in plants

- *Limitation of uptake/transport* of salts (by synthesis of organic substances to raise internal osmotic potential)

  > e.g. proline (up to 30 per cent of amino acid content in some salt marsh plants)
  >> mannitol (carbohydrate) used by brown seaweeds
  >> glycerol: in halophilic phytoplankton *Dunaliella*

- Unlimited uptake but salts *compartmented* → structures less susceptible to toxicity effects *or good tolerance mechanisms* for high osmotic potential in cells
- Control of internal concentration of salts and ion balance, by *excretion* of salts (e.g. salt glands on leaves – modified stomata: *Limonium*, *Spartina*)
- *Selective* ion uptake and transport at root or organelle surfaces

Effects of stress on animal populations in stressed ecosystems

Stressed ecosystems for animals are usually those in which temperature or other climatic conditions are hostile to survival: key metabolic processes, such as respiratory activity, reproductive activities or behavioural activities, may all be impaired if, for example, the external temperature is above or below the animal's tolerance range.

To take one example, **poikilothermic** (cold-blooded) desert animals such as lizards pay a lot of attention (and a commensurate energy price) to ensure that they always occupy the optimal part of their immediate habitat in temperature terms, basking in sunlit areas if their internal temperature declines, and moving to the shade as they heat up. By contrast desert **homiotherms** (warm-blooded animals) of similar size to lizards display relative indifference to diurnal ('24-hour') variations in temperature. Large temperature fluctuations are needed before they respond by moving to warmer or cooler parts of their habitat. Mad dogs and Englishmen are not the only mammals which go out in the midday sun. An extreme example is the antelope ground squirrel (*Ammospermophilus harrisi*) of the Sonoran Desert in the south-western USA, which continues to be active even in air temperatures of 43°C, and when the sand beneath its feet can be as hot as 66°C. Lizards have problems of overheating if they venture out in the sun in such conditions, but the squirrel has evolved physiological adaptations to minimise water loss and maximize heat loss (its urine is almost solid, for example, and it can radiate heat from its body surface incredibly quickly). It also has a range of behavioural adaptations to cool itself down (for example, drooling saliva on to its paws, then washing its head to wet its fur, so increasing evaporative heat loss).

The net energy costs paid by homiotherms to obtain such benefits are very high (compared with lizards, mammals have to eat huge amounts of food just to maintain their internal constant temperature) but the rewards are also enormous in terms of ability to cope with varying and high-stress environmental conditions. Natural selection has ensured the success of the genetic traits which confer warm-bloodedness, and its consequent advantages for stress tolerance in animals.

Strategies for adaptation in stressed ecosystems

If the intensity of environmental stress is high enough no life can exist, and no functioning ecosystem can exist either. However, extremely simple ecosystems can occur in conditions which we might imagine would be impossible to sustain life (see Case Study 4).

Organisms which (by natural selection processes) have successfully evolved the necessary traits needed to survive intense stress pressures (Box 5.3) can exploit 'difficult' stressed habitats, often with considerable success in terms of population size and distribution. After all, harsh, high-stress ecosystems cover a large percentage of the Earth's surface (hot deserts, for example, cover about 20 per cent of the land area of Earth; cold deserts like Antarctica nearly another 15 per cent) so there is ample scope for colonisation and spread of suitably adapted species in these areas. The extreme stress associated with ice-covered habitats is usually thought virtually to preclude plant growth. However, this is not entirely so. Beneath floating ice in the oceans there exists an ecosystem about which we still know relatively little, but which seems to play an important refuge role in the overwinter functioning of the Arctic and Antarctic ecosystems (Campbell 1992). **Sympagic** ecosystems of this type occur where phytoplankton become locked up within the ice as it forms (sometimes reaching very high densities: up to 33 million algal cells per litre of ice) or form dense overwintering aggregations immediately beneath the ice. Some sixty species of diatoms are known to be tolerant of freezing in the sympagic realm. The sympagic algal blooms on and in the bottom of the pack ice are now thought to be a very important resource, grazed by krill and other zooplankton (especially copepods) during the winter months, and contributing as much as 12 per cent of the total primary productivity of the ice-covered oceans.

Stress-tolerant species pay a severe price for their adaptations in terms of a reduced ability to survive in environments where the stress does not exist, or exists only at a

Case study 4

One of the best examples of a simplified ecosystem, occurring under extreme stress conditions, is to be found in the Antarctic dry valleys. The dry valleys occupy some 5,000 km^2 of South Victoria Land, in that otherwise ice-clad continent (Campbell 1992). These ice-free valleys are kept that way by their fringing mountains, which prevent the movement of wet air from the ice-cap. Here we find an endolithic ecosystem, occurring to a depth of a few millimetres within the interstices of the translucent dry rocks (such as quartz) of which the valley walls are constructed. A few lichens, plus cyanobacteria, and associated fungal and bacterial decomposer organisms, make up what is arguably the simplest functioning ecosystem that we know about to date. The cold and lack of water prevent the occurrence of any animals to exploit the tiny primary production of this ecosystem, perhaps the most highly stressed ecosystem in the earth's biosphere. It is interesting to note that conditions in the Antarctic endolithic ecosystem are only a little better than those prevailing, so far as we know from investigations there to date, in one of the two other feasible biospheres within the solar system – Mars (the other possible biosphere is on Europa, one of the moons of Jupiter: see Chapter 3). In terms of temperature and water regime an Antarctic endolithic ecosystem, like that of the appropriately named Mars Valley in Antarctica, wins by only a slight margin over a comparable location on Mars (the main difference is that there is less oxygen available in the Martian atmosphere). Such localities may be excellent places to search for life on Mars, if any does indeed exist (see also Chapter 3).

Box 5.3

Pressures on plant survival in a stressed ecosystem

Environmental stress *high*, disturbance *low* (e.g. cold temperatures in arctic-alpine mountain ecosystems)

↓

photosynthetic production limited

↓

competition much less important: less crowding

↓

to survive, plant needs stress-tolerance traits

- S-strategists often small, slow-growing, protect tissues from worst effects of stress (e.g. tundra plants: up to 90 per cent of plant biomass underground)
- Specialist protective growth forms: e.g. 'cushion' plants of high mountains: (e.g. moss campion, *Silene acaulis*)

reduced level. This is because the resources which the organism has poured into building the necessary stress-tolerance structures, processes, behaviour, or whatever the adaptations may be, are normally more or less useless in helping it compete with less heavily adapted organisms when conditions for survival are better. Stress-tolerant organisms are poor competitors outwith their stressed habitat, but compete extremely well with non-stress-adapted species where (or when) the particular stress conditions to which they are adapted are actively influencing the habitat.

Two examples illustrate this. The emperor penguin (*Aptenodytes forsteri*) is undoubtedly the best adapted of any warm-blooded animal to cold stress (except for *Homo sapiens*, but we cheat by using things like duvet jackets and nuclear reactors to make conditions more to our liking, for example, in the US base at the South Pole). The emperor penguin is the only animal capable of surviving the perpetual cold and darkness of the Antarctic winter on land, where temperatures may drop as low as −30°C (even colder when wind chill factors are taken into account) for days or weeks on end. Male emperor penguins not only cope with two months fasting on the ice-cap but also successfully incubate their eggs (which lie on their feet, tucked under the feathers of the bird's lower abdomen) as they huddle together in large groups for the duration of the terrible winter conditions of Antarctica.

Emperor penguins are a successful species. They have found and exploited a cold-stressed environment where there are effectively no competitors and no predators for the duration of the most vulnerable period of their life cycle (i.e. while they incubate their eggs and look after their young chicks; however, high mortalities of chicks do sometimes occur due to exposure and starvation during prolonged blizzards). At least twenty emperor penguin rookeries are known, scattered right around the Antarctic coastline, and supporting over 350,000 birds. The price these birds pay for their success in coping with cold stress is to be absolutely prevented, by the very adaptations which allow them to live here, from expanding their area of colonisation beyond the coasts and neighbouring seas of Antarctica. They never venture further north than Tierra del Fuego, at the extreme southern tip of South America. Other penguin species (e.g. Magellanic penguins, *Spheniscus magellanicus*) which are less strongly adapted to cold conditions, travel as far north as the coasts of Brazil.

Even at first glance (Plates 5a, 5b) the morphological differences between the two species make it obvious which is the better adapted to cold stress. The emperor penguin is much bigger (a lower surface area:volume ratio makes it easier to retain heat) and much heavier than its Magellanic cousin. Emperor penguins have a sleeker, more bulky appearance due to their denser coat of feathers and the thicker layer of blubber. On land they waddle along slowly, conserving energy, while the Magellanic penguin can run quickly, climb rocks and seems altogether much less concerned about expending energy.

The Magellanic penguin has put less effort into developing expensive specialist stress-tolerance traits than the emperor penguin, and as a result it can compete successfully with other fish-eating birds and mammals, even into the warmer waters of the mid-Atlantic. But Magellanic penguins would not have a hope of surviving the Antarctic winter on land. Instead of having a reasonably safe (though chilly) haven for their eggs, they suffer a high loss rate of eggs and chicks due to fierce predation – from foxes, rats, gulls and skuas, for example, the great skua (*Catharacta skua*) of the South Atlantic – in the rookeries which they occupy on the shores of Argentinian Patagonia.

How do plants cope with high-stress ecosystem conditions? A good example (perhaps better termed 'stress avoidance' in this case) is that of the bluebell (*Hyacinthoides non-scriptus*), a common woodland plant of Northern Europe (Figure 5.1). Bluebells are famous for the beautiful carpets of blue flowers they produce in early spring in oak and other deciduous woodlands. What the plants are doing is to exploit, very successfully, a brief window of opportunity between the cold, dark days of winter (when conditions are too stressful to allow the plants to survive above ground) and the warm, sunny days of summer (when, unfortunately for low-growing plants like bluebells, conditions would be perfect for growth except for the fact that the trees have meanwhile built a thick

Plate 5 (a) Emperor penguin (*Aptenodytes forsteri*)

Original photo: Glasgow University, with permission

(b) Magellanic penguin (*Spheniscus magellanicus*)

Original photo: K.J. Murphy

Figure 5.1 Bluebell (*Hyacinthoides non-scriptus*)

canopy of leaves above them, which imposes a lethally severe shade stress on any plant trying to live on the ground beneath the trees). To survive this shade stress, and to continue to occupy the woodland environment successfully, bluebells have evolved a range of adaptive traits which collectively permit them to live here. They have big underground storage organs (**bulbs**), stuffed with sugars in the form of starch, which provide them with the means to grow quickly once the spring weather gets warm and bright enough for above-ground growth to start. Given this head start they are then in a race against time (and their bigger neighbours) to grow their foliage, then use it to capture enough light energy to allow them to flower, set seed and (the vital step) replenish the starch reserves in their bulbs, before the death sentence imposed on the adult plants by the developing tree leaf canopy above them is carried out. They have about two months maximum to get through the whole of the above-ground (established phase) part of their life cycle before they retreat back to the underground regenerative phase, resting in the form of bulbs and seed until the following spring. Not many plant species have successfully evolved the right set of traits to succeed in heavily shade-stressed habitats like an oak woodland. This is one reason why bluebells are such a dominant feature of North European oak forest ecosystems; there are few competing species to grow in their midst, and the **phenological niche** they have occupied is (almost) all theirs to exploit.

Like the emperor penguin, the price paid by the bluebell is quite high. Under summer conditions the plants have no chance of competing successfully with other plant species which do not have the inbuilt disadvantages of the bluebell. Effectively they are therefore confined to their woodland stronghold, and this is a diminishing retreat as the area of deciduous woodland in Europe steadily declines. The bluebell's fate is irretrievably linked to the fate of its 'host' habitat by its very possession of the stress-tolerance traits which allow it to exploit that habitat.

These two examples illustrate the main principles of adaptation shown by organisms which live in environmentally stressed ecosystems. The CSR theory and r–K models (see Chapter 2) both suggest that we can identify broad categories of adaptation for plant and animal communities living in stressed ecosystems. These adaptations can be anatomical, physiological or behavioural (usually, in fact, combinations of all of these: as we saw in the case of the emperor penguin).

Plant tolerance of high-stress ecosystem conditions

Terrestrial and aquatic forests: stress produced by shade

Highly productive forests, both in terrestrial environments and in the seas, paradoxically also produce heavily shaded ecological habitats which are distinctive and significant components of the whole ecosystem.

Plants which, unlike the bluebell, have not managed to find a way of evading the problems of surviving shade stress in forest ecosystems, have evolved a range of mechanisms to maximise photosynthetic gain in the low-energy habitats which occur beneath the canopy in forest ecosystems (both on land and in underwater kelp forests: notably the giant kelps, *Macrocystis*, which form dense seaweed forests off the Pacific coast of North America). These mechanisms include:

- reduced respiration (this lowers the compensation point, resulting in an 'energy profit' even under low light conditions)
- increased unit leaf rate (= higher photosynthetic rate/unit energy/unit leaf area)
- increased chlorophyll per unit leaf weight
- thin leaves (only a few cells thick – so as much of the chlorophyll in the cells is as close to the leaf surface as possible), with the leaves often arranged to minimise self-shading.

In underwater ecosystems there are additional problems for plants (over and above the shade effects produced by canopy absorption of light). The water selectively absorbs red wavelengths of light more than blue or green light, thereby changing the quality of light which penetrates the kelp forest canopy. There is also an exponential decrease in total photosynthetically-active radiation (PAR) energy with depth, because the water molecules absorb light energy efficiently (see Figure 5.2). This can be measured as a PAR attenuation coefficient: the slope of line relating $\log_e$ PAR intensity to depth beneath the water surface. The compensation depth is the depth where the plants' compensation point is reached; here photosynthetic carbon gain just balances respiratory carbon demand in plants. The compensation depth is much closer to the surface under a kelp forest canopy than in open sea water (though even there seaweeds rarely occur deeper than about 20 m below MSL).

The underwater forest canopy of kelps can be substantial. Kelps are large brown seaweeds (of the Phaeophyta), the biggest of which, *Macrocystis*, can have fronds up to 50 m long and form permanent beds large enough to be marked on charts as a hazard to shipping. Even the smaller kelps, such as *Laminaria* (which grows around the coasts

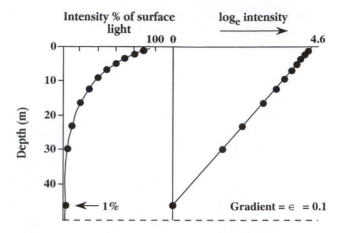

Figure 5.2 Curves showing absorption of light with increasing depth underwater, as percentage of surface light intensity (left), and as $\log_e$ transformed surface light intensity (right). The point labelled 1 per cent is usually considered to be the compensation depth for phytoplankton survival (see text for details)

of the British Isles) may be 2–3 m long, with broad laminae, which float up in the water when submerged at high tide to form a dense, tangled canopy of photosynthetic tissue. Other seaweeds growing under this canopy may be either epilithic (growing attached to the rock) or epiphytic (growing on the stipes (stems) of the kelp plants). In both cases it is red algae (of the Rhodophyta) which are particularly common here (e.g. *Lithothamnion*, *Ptilota*, *Membranoptera* under a *Laminaria* canopy). These plants show several of the standard shade adaptations seen in terrestrial shade-tolerant plants (e.g. delicate, thin photosynthetic structures), but in addition their possession of red photosynthetic pigments is a decided advantage because these pigments are ideally adapted to capture the remaining quanta of blue-green light which filter through the water and kelp canopy above them.

Plant survival in drought-stressed ecosystems

Relatively few families of plants have managed to evolve the necessary traits needed to survive the very hostile conditions typical of arid and semi-arid ecosystems (e.g. desert and semi-desert habitats). Adaptations to the stress caused by shortage of water include succulence (inflated stem or leaf tissues holding a reservoir of water); small, leathery leaves (to minimise loss of water during photosynthesis); and often deep or widespreading root systems. Plants containing lots of water are attractive to grazing animals, so many plants in arid habitats have armed stems and leaves (e.g. thorns, spines) to discourage animals from eating their tissues.

Succulence is an excellent example of a functional trait for survival of drought stress which has probably evolved more than once during plant evolutionary history. The Cactaceae are the best-known examples of succulents. They occupy a wide range of drought-stressed ecosystems. They are superbly effective tolerators of drought stress. There are about 2,000 species of cactus, in 140 genera, spread through the dry parts of (mainly) the Americas. Clearly their adaptations have been a major success story in evolutionary terms. The most primitive tropical cacti species have fairly normal-looking

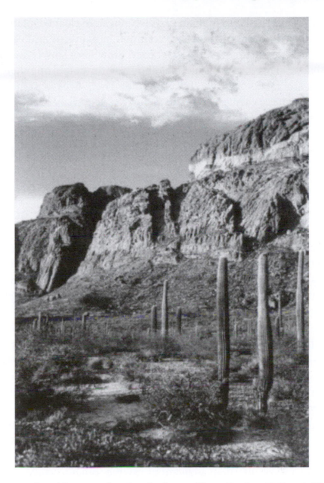

Plate 6 Saguaro cactus (*Cereus giganteus*): Organ Pipes Cactus National Monument Area, Arizona, USA

Original photo: K.J. Murphy

leaves. However, in the hottest and driest habitats, cacti such as the giant saguaro (*Cereus giganteus*, the biggest cactus, growing up to 15 m tall: Plate 6) have evolved remarkable adaptations to cope with water stress. These include the best examples of stem succulence of any plant. In the process they have lost their leaves entirely (leaves being far too profligate losers of water for comfort in these sorts of ecosystem conditions). Their chlorophyll is, instead, in the surface (epidermal) layers of the enormous barrel-shaped inflated stems which constitute the plant's water reserve. Species of *Cereus* and other cacti (e.g. cholla (*Opuntia*)) found in hot, dry desert ecosystems like the Sonoran Desert of Arizona have shallow root systems. These are designed to catch the water which arrives only intermittently from showers of rain, and which evaporates or runs off before it can penetrate deep into the soil. To save on water loss from within the plant, the pores (stomata) in the stem epidermis open only during the cooler night to allow entry of carbon dioxide into the plant's cells for photosynthetic fixation. However, carbon fixation can happen only if light is present. So the cacti have evolved a bit of physiological trickery to allow the carbon dioxide to be stored chemically inside the cells until it is needed during the day. Such physiological adaptations to drought stress (so-called **CAM photosynthesis**) are common in plants of arid ecosystems.

Like many stress tolerators, the giant cacti grow very slowly and are highly intolerant of disturbance. This is part of the price they pay for colonising their high-stress habitat. A big saguaro cactus may be 200 years old. Their wicked spines deter most grazing animals, and a large cactus is immune to grazing disturbance. However, lightning strikes or fires can kill the biggest specimen, and they are also badly damaged by fools who use them for target practice (though completely illegal, it is not uncommon to see the scars of bullets or shotgun blasts on cacti in the western USA, even in protected areas like the Organ Pipes National Monument Area of Arizona).

Plant adaptations to nutrient stress in low-fertility conditions

Plants which have successfully colonised habitats where nutrients (e.g. nitrogen) needed for growth are in short supply usually have only two basic strategic options for coping with the shortage of nutrients in the soil. Either they tolerate the lack of nutrients by being slow-growing and usually small, or they are larger, faster growing and have got around the problem by finding new sources of nutrients.

The carnivorous plants are an excellent example of a functional group which has followed the latter strategy. They show a wide range of adaptations to catch insects and other small animals, kill them, and absorb nutrients released from their decaying bodies. Deficiency of nitrogen is particularly dealt with by such adaptations, given the N-rich content of animal remains. Carnivory in plants is a functional trait which has possibly only a single phylogenetic origin. All carnivorous species are in the same part of the plant phylogenetic tree (ranunculids to asterids: Chase *et al*. 1993) so the original mutation(s) producing the ancestral carnivore probably occurred some way back, beyond the division separating ranunculids off from asterids/rosids. However, the sarracenid pitcher plants have rather different adaptations to the other carnivorous species, and it is possible that carnivory in this group evolved as the result of a separate mutation event, occurring independently from the evolution of carnivory in the other groups.

In comparison with the carnivorous plants, other functional groups of plants adapted to low-nutrient conditions are much more phylogenetically variable. In Chapter 2 we introduced the isoetids (see Box 2.1) and these provide an excellent example of a stress-tolerant plant functional group found in low-nutrient conditions. The trait-set which these plants have in common includes several adaptations to life in nutrient-poor ecosystem conditions. These include the morphological root adaptations (mentioned in Chapter 2) to assist in foraging for dissolved carbon dioxide in the interstitial water of the sediments in the lake ecosystems where isoetids occur. The plants also tend to have a low biomass turnover rate, a high root:shoot ratio, and slow growth rates. Most are perennials, with a tendency to rely on asexual reproduction rather than seed dispersal to produce their offspring. There are exceptions: awlwort (*Subularia aquatica*) is an annual isoetid found in Scottish lochs, which relies absolutely on seed production to produce its next generation; even very small *Subularia* plants, in the most hostile of conditions, will typically manage to produce a few seeds, such is the priority given to seed production in annual plants.

The adaptations seen in isoetids are all designed to maximise their chances of survival in nutrient-stressed ecosystems, with only (at worst) moderate intensities of disturbance (Farmer and Spence 1986; Murphy *et al*. 1990; Murphy 2002). Oligotrophic lakes offer perfect examples of this combination of conditions: look for isoetids there and you are unlikely to be disappointed.

The role of environmental stress in countering bio-invasions

Invasion of ecosystems by non-native species ('bio-invasion') is a major threat to biodiversity worldwide. Just how prone to invasion different ecosystem types are, and what makes an ecosystem more or less open to invasion, is the subject of much current research (e.g. Burke and Grime 1996; Li and Norland 2001). The intensity of stress affecting an ecosystem may be one important factor. An example is provided by data from low-nutrient urban bushland ecosystems in Sydney, Australia (Lake and Leishman 2004), heavily invaded by exotic species (with fifty-seven exotic species recorded out of a total flora of 133 species present in the area). These data suggest that increased soil fertility at locations downstream of urban stormwater runoff points, which produced a relaxation of the stress caused by nutrient shortage in the soil, may play a significant role in how open these systems are to invading plant species. Low disturbance, high stress sites, with little or no grazing (primarily insect herbivory) or damage by human trampling (away from tracks and firebreaks), and lacking any nutrient enrichment, were consistently free of exotic species and had a high species richness: averaging 25.3 species per 400 m^2. Physically disturbed sites (close to tracks or firebreaks, or with greater insect damage), but with low-nutrient soils, and consequently similar stress levels to the first group, had low numbers of invasive plants (usually only one species: a grass, *Andropogon virginicus*). However, the biggest difference was seen at the locations with nutrient-enriched soils (i.e. reduced stress) where not only were there large numbers of exotic species, but also significantly lower native species richness (averaging only 9 species per 400 m^2). Relaxation of the intensity of environmental stress appears to be a prerequisite for invasion.

Stressed ecosystems: some conclusions

In this chapter we have shown how stress-tolerant organisms must pay an ecological price for the adaptations which allow them to succeed in occupying high-stress ecosystems. The precise cause of the physiological stress which the biota of such ecosystems experience is much less important than the outcome as far as it affects the organisms concerned. In practical terms, a penguin, which is prevented from colonising warmer waters because of its high blubber investment, experiences the same sort of potential competitive disadvantage (in lower stress conditions) as a woodland or sublittoral plant which has invested heavily in shade-tolerance adaptations. If either of these organisms attempts to colonise habitats which do not experience the high intensities of the relevant environmental stress, they are very likely to fail in the face of competition from more productive and faster growing organisms which have not invested in the relevant stress-tolerance adaptations. Equally, organisms which do not have the right stress-tolerant adaptations are highly unlikely to succeed in high-stress ecosystems. An important result is that stressed ecosystems (and many of them are very extensive) offer a refuge to specialist organisms, resulting in an overall higher biodiversity (i.e. across the range of high to low stress ecosystems) than would be the case if the specialist stress-tolerant species did not exist. In addition, there is some evidence that stressed ecosystems may be naturally resistant to invasion by exotic species, which has a tendency to reduce ecosystem biodiversity.

Although we may think of stressed ecosystems as 'difficult' places for survival, and though such ecosystems may have (in many cases) lower biodiversity than other types of ecosystems, it is important to realise that the sum total of biodiversity supported by the biosphere of the Earth would be greatly diminished if the stressed ecosystem biota were lost. One big problem here is that, by their very nature, the biota of stressed ecosystems tend to live perilously close to the edge of what is survivable. If human activities increase these pressures beyond the point of tolerance (see Chapter 2), there is a strong risk of extinction for the populations which occupy the stressed habitat. Desertification problems are a good example: see Chapter 10 for more on this.

Summary

- This chapter discusses the characteristics of stressed ecosystems and some of the principal characteristics of the organisms which specialise in living in these challenging conditions.
- Cold, searing heat, water shortage, lack of nutrients, and/or the presence of potentially toxic materials in the ecosystem all produce stress-tolerance responses in the animal, plant and microbial organisms which are adapted to these conditions.
- The investment of resources in such defences against environmental stress (which include physiological, morphological and behavioural adaptations) tends universally to exclude these stress-tolerator organisms from ecosystems with better conditions, where less heavily adapted species can more effectively outcompete or predate the stress tolerators.
- Stress-tolerant species are generally locked into their chosen ecosystem conditions by the very existence of the adaptations they have evolved to combat the effects of stress.
- The more heavily adapted the species (i.e. to more extreme environmental conditions) the more this is so; examples are given (e.g. penguin species) which illustrate this ecological phenomenon.
- Stressed ecosystems show signs of being naturally hostile environments for potential invaders: this may be a factor of importance in helping to maintain the biodiversity of such ecosystems.

Discussion questions

1 Some ecosystems qualify in their entirety as experiencing high-stress conditions. In others only some of the habitats within the ecosystem offer stressed conditions to the biota living there. Is it possible to draw up a list of different ecosystem types which would fall at different points along a gradient of stress: from 'whole ecosystem stressed' to 'only a few stressed habitats within the ecosystem' (hint: read the section on intermediate habitats in Chapter 7)?

2 What are the most important sources of stress conditions for (a) plants and (b) animals in land and water ecosystems, on a planet-wide basis?

3 Can we really describe an organism as 'successful' if it is confined to only a narrow band of environmental conditions as a result of its evolving stress-tolerant adaptations?

Further Reading

See also

Definitions and examples of stress tolerators, Chapter 2
Disturbed ecosystems, Chapter 6
Competitive and intermediate ecosystems, Chapter 7

Further reading in Routledge Introductions to Environment Series

Biodiversity and Conservation
Environmental Biology
Natural Environmental Change
Oceanic Systems

General further reading

The Crystal Desert. D.G. Campbell. 1992. Martin, Secker and Warburg, London.

A superb and entertaining description of Antarctic ecosystems and the people who have lived and studied in Antarctica, from the explorers, whalers and sealers of old to modern-day scientists. Includes a vivid and moving description of the carnage wreaked on the great whales by human exploitation. Highly recommended reading.

Studies in Plant Survival. R.M.M. Crawford. 1989. Blackwell, Oxford.

A useful description of plant responses to stress (and to disturbance) in the form of a series of case studies of different ecosystem conditions, from the forest floor to tundra.

6 The role of disturbance and succession in ecosystem functioning

The abilities to endure unstable conditions and to recover quickly from events that destroy either the habitat or the organisms themselves are the hallmarks of disturbance-tolerant plants, animals and micro-organisms. Disturbance tolerators tend to colonise early in the successional recovery process that follows environmental catastrophes, large or small (from trampling damage on an upland path to volcanic eruptions). Lower-level intensities of disturbance are a commonplace feature of many ecosystems. Successional changes following disturbance events play an important role in the functioning of disturbed ecosystems. This chapter covers:

- **Defining disturbance**
- **Succession: community change over time in ecosystems**
- **Colonisation of lifeless surfaces**
- **Land–sea interface**
- **Ecosystem resilience and fragility**

Defining disturbance

Ecosystems exist in a constantly changing world. Change results from processes in the abiotic or physical environment, from changes in the biotic environment, the living ecosystem community and by human actions. **Disturbance** is both a natural and normal part of the environment, but is often increased in scale and accelerated in effect by human impacts. Disturbance is defined as any influence on an ecosystem, which increases the probability of destruction of biomass of the organisms present (see Box 6.1). In disturbed ecosystems these influences may be either biotic or abiotic. For example, grazing is a strong disturbance pressure on plants in a prairie ecosystem, but so are abiotic disturbances (e.g. caused by lightning-induced grass fires).

Disturbance events affecting ecosystems range from the intermittent and cataclysmic (e.g. a volcanic eruption: see Chapter 2), to more permanent and lower intensity pressures, such as the constant effects of disturbance produced by wave action and grazing on marine rocky shore ecosystems.

Disturbance affects different ecosystems in different ways. Some species are highly adapted to tolerate disturbance. Disturbance-tolerance adaptations in plants include rapid life cycle, coating of trunks with bark, which is resistant to the high temperatures associated with burning, and rapid regrowth of tissues from intercalary ('protected') meristems in response to grazing. The response of a plant community to disturbance will thus be a function of the nature of the disturbance and of the response of the members of that community to particular disturbances. As changes in the biotic and abiotic environment are normal, change in response to *natural* environmental change is part of the

Box 6.1

Disturbance: general principles

Stress *low*, disturbance *high*: big problem is damage to the plant's
biomass – either partial or total destruction
↓
e.g. grazing usually causes only partial destruction; landslide on unstable
mountain scree slope may completely destroy plants
↓
plants rely heavily on regenerative phase (seeds, spores, vegetative propagules)
+ protective structures

- Life in unstable conditions of an ecosystem prone to disturbance needs different traits from C or S-strategists
- R-strategists (first described from disturbed 'ruderal' habitats, along tracks and roadsides): where disturbance is common
- Traits: e.g. protected meristem, as in grasses (most successful group of grazing-tolerant plants) have 'growing point' of the plant placed near to the soil → rapid regrowth after foliage grazed off
- Fast-growing, get through their established-phase cycle quickly to produce regenerative propagules (main insurance policy against extinction)
- Put a lot of effort into seed or other propagule production, and dispersal

behaviour of ecosystems. Where disturbance is small or cyclical, for example, seasonal or yearly changes in climatic conditions, the response will also be relatively minor. But there are also directional changes which result from what we have identified as natural disturbance. Some of these relate to long-term changes in climate, or to major geomorphological processes. Others are related to changes that have been brought about by modification of the environment by plant communities. This results in further, directional change to the plant community itself in the process known as 'succession'. This important concept was introduced in Chapter 1, when its role in the development of ecological science was outlined. It is analysed in more detail in the following section.

Succession: community change over time in ecosystems

The concept of succession is over a century old. Although it has been much modified and argued over by ecologists, it still provides a useful way of understanding the nature of the dynamic reciprocal relationships between plant communities and their environment.

The basic principles of succession are outlined in Box 6.2. Starting with a surface devoid of vegetation, plants specifically adapted to colonise such harsh environments begin an environmental alteration process which culminates in the development of a relatively stable plant community, which will persist and may have high biological productivity (though this of course depends in part on other factors: we would not expect an Arctic community ever to reach the productivity of a tropical rainforest, no matter how long we waited). At successive stages or seres, the plant community, and higher

Box 6.2

Stages in a typical plant succession

1 Initiation

The starting point of any succession is a bare surface. It may be 'new', e.g. an emergent shoreline, or more commonly a surface stripped of any previous vegetation cover by natural or human agencies.

2 Colonisation (Sere 1)

The first plant growth is based on a small number of specialised, highly stress-tolerant plant species. Total biomass is low, and soil is rudimentary, generally lacking organic matter and balanced available nutrients. Typical colonisers are bryophytes, and vascular plants with tolerance of extreme water and nutrient status conditions (either high, low or alternating).

3 Development (Sere 2)

As soil conditions improve, highly stress tolerant species are replaced by more productive and competitive species. Productivity increases and soil biology develops. Typical species in this sere include grasses and weeds. Both of these are quite tolerant of disturbance which is often a feature of the developmental sere, in which substrate conditions may remain unstable and alternation between different environmental conditions may occur.

4 Mature (Sere 3)

By this point the ecosystem has developed to the extent that vegetation cover is dominated by competitive species, though not necessarily those with a very long life cycle. Soil conditions are stable, and nutrient and water conditions are not major problems in the ecosystem. Typical species are competitive grasses, bushes and smaller trees. Non-vascular plants are minor components, and the range of higher trophic and decomposer species is considerable.

5 Climax (Sere 4)

The final stage sees the development of a vegetation cover which is relatively stable and persistent. It is often dominated by large trees, with a long life cycle. There is little or no evidence of the initial abiotic or biotic environmental conditions of the area which exist at the beginning of the successional sequence. The issue of whether or not there is such a condition as stable climax is controversial.

trophic structure, tends to become more complex. Stress and disturbance-tolerant species are progressively replaced by species with high competitive ability as the environment becomes less stressful and more stable. Important changes which take place are the development of soil conditions in which nutrients and water are more freely available. These circumstances are largely the product of the accumulation of organic matter in the soil, the development of humus and the cycling of nutrients by soil decomposers. All of these changes are dependent on the development of vegetation cover. The end-point, the development of a stable vegetation cover, which generally has higher primary productivity than preceding seres, and supports a more complex ecosystem than the earlier stages, was termed 'the climax' by Clements (1928), one of the proponents of the theory of succession. The simple notion of climax, certainly that solely determined by climate, as was proposed by Clements, is now qualified by ecologists. However, the general structure of successional development still has validity, and helps us understand how plants not only respond to disturbance, but are also responsible for the creation of disturbance to themselves. The way in which plant communities, and the animal communities which depend upon them at higher trophic levels in their ecosystems, respond to various types of disturbance through succession and other ecological responses is considered in the following examples.

The colonisation of lifeless surfaces

Scree

Scree is an accumulation of primarily angular material at the base of an exposed cliff. It will pile up to form a sloping accumulation of freshly eroded material, to a maximum angle of about 36°. This is known as the angle of repose, which varies somewhat according to the nature of the material deposited. If material is deposited at a slope greater than this angle, slopes become unstable, and material will move to an equilibrium angle of slope less than 36°. Once the scree slope is stable, vegetation will begin to colonise the new surface. Mosses such as woolly-fringe moss (*Rhacomitrium*) can exist on the rock fragments, even though this surface has no water-retention ability and virtually no available nutrients. As accumulations of dead moss build up in the interstices between the rock shards, vascular plants such as thrift (*Armeria maritima*), which are drought tolerant and can withstand severe exposure, can colonise the scree slope. An example of this is shown

in Plate 7. Further succession will allow various heather (*Erica*) and rush (*Juncus*) species to become dominant. By this stage most of the rock surface is vegetated, and a thin, peaty substrate is developing. Given the harshness of this environment, which may be located in a climatic zone with a long winter which has many freeze–thaw cycles and experience frequent exposure to strong winds, colonising vegetation must tolerate conditions of physiological drought and extreme nutrient deficiency. In addition, frost action may cause disturbance to the substrate, and plants must be able to endure such action occurring each winter. At the final stage heath or scrub woodland will develop. In most cases in Britain, the actual final stage is dependent upon human impact and management in the area, and is often an anthropogenically maintained sub-climax, rather than a true climax community.

Plate 7 Vegetation colonising a scree slope on the island of Rum, Scotland

Original photo: G. Dickinson

Ice margins and permafrost conditions

Surfaces are exposed at the margins of retreating ice-caps. Advance and retreat of continental ice-caps is a normal environmental condition associated with long-term changes in global climate. The last time that the ice sheets were in a growing phase ended about 12,000 years ago, though minor variations have occurred during the past millennium. The effects of human-induced global climatic change upon the ice bodies of the planet may be important, though as yet we cannot predict exactly what the outcome may be. For example, though global warming might be expected to cause ice melting, this may not be the case. Increased precipitation in polar regions, which are currently cold deserts, may actually increase ice cover. Changes in the global ice budget are highly significant, since ice comprises about 2 per cent of all water on the planet, several times more than the combined totals of fresh and atmospheric water. Quite small changes in the total volume of ice will cover extensive areas of land when glaciers are expanding, or reveal land surfaces when the volume of ice is decreasing.

There is clear evidence that glacier advance and retreat have occurred many times naturally in the geological past. The current climatic conditions are a period of relatively warm global conditions in a sequence of cold and warm periods. These climatic variations are complex in origin, and appear related to minor variations in the Earth's orbit, and to changes in factors, which influence the receipt of solar radiation (Mannion 1991). Exposure of new land surfaces following glacial retreat has been a common and natural process over the past million years, particularly affecting the high-latitude land masses of the northern hemisphere. Development of vegetation on such surfaces is an example of natural, primary succession. The pattern of succession is similar to that on scree slopes, except the species involved are tolerant of very cold conditions and unstable substrate.

The land surface exposed by a retreating ice sheet is nearly sterile. This surface is composed of different sorts of materials. These include sorted, unconsolidated sediments such as sand and gravel deposited from flowing glacial melt-water, and finer silt and clay, which have been deposited in still water conditions from temporary lakes at ice margins. There are also likely to be areas of unsorted till which has been deposited directly from the ice. Areas from which all superficial material has been stripped, exposing solid rock at the surface, will also be exposed. Thus the initial substrate conditions vary considerably. Nutrient supply and water conditions will vary over short distances. Continued disturbance is likely to occur as a result of cold climatic conditions, which will cause **cryoturbation** of substrate materials. Below the ground, material will be permanently frozen as a result of the long period of contact with the overlying ice body. This is termed **permafrost**, and it underlies much of the Earth's surface in the high latitudes. Figure 6.1 shows the distribution of permafrost in the northern hemisphere. Much of this is a relic of the last major advance of the continental ice sheets, and thus has persisted in some areas for several thousand years. Permafrost has a profound influence on ecosystem development. Only the uppermost part, rarely more than one metre in depth, thaws out temporarily in the short summer. The areas at the margins of glaciers have cold climates, characterised by long winter periods during which temperature rarely rises above 0°C and the surface is normally snow-covered. All plant growth must be concentrated in a period typically less than ninety days in duration. The upper part of substrate is composed of mobile sediment overlying permanently frozen material.

The substrate – soil is not really an inappropriate term in the early stages of development – is mobile. Sediment is moved by water and wind, as well as by cryoturbation, which churns up the upper part of the surface zone. Drainage is poor and changes rapidly.

To this pattern of naturally occurring disturbance must be added human impacts. Economic development, for example, for minerals such as oil, causes great disturbance to this fragile ecosystem. Human constructions such as roads and buildings disrupt permafrost by causing deeper melting. The whole surface zone can become so unstable that plant cover is eliminated, and the land becomes a swamp in summer, pitted with

Figure 6.1 Distribution of permafrost in the Northern Hemisphere

water-filled pools, and an ice desert in winter. This type of surface has been termed 'thermokarst'. Even minor human impacts can have serious outcomes for ecosystems close to permanent ice.

In such hostile conditions, only a few plant species can survive. Environmental conditions are perilously close to the uninhabitable high-stress plus high-disturbance combination (see Chapter 2). Species able to cope with the combination of cold stress and disturbance include bryophytes and lichens, which are generally the first colonisers. Grasses and sedges develop at later stages in succession, as soil conditions improve through the incorporation of organic matter and humus in the developing profile.

Landform facet:	stream bed	recent meander deposits	older fluvial deposits	glacial drift
Substrate	gravel	sand and silt	sand	diamicton
Permafrost	> 5 m	5 m	discontinuous < 5 m	continuous < 5 m
Vegetation type	none	grasses, reeds and moss	scrub wood	coniferous forest
Vegetation age	n.a.	< 50 years	50–100 years	> 100 years
Vegetaion height	n.a.	< 1 m	< 10 m	> 10 m

Figure 6.2 Vegetation in a typical Arctic area partly underlain by permafrost

Climatic conditions and permafrost hinder tree growth. A few shallow-rooted trees, such as black spruce (*Picea mariana*) in North America and dahurian larch (*Larix gmelinii*) in Siberia, can avoid the worst of the freezing conditions in the soil by keeping their root tissues in the warmer deposits close to the surface. Otherwise the climax vegetation is usually tundra, dominated by dwarf shrubs, herbs, lichens and mosses. Figure 6.2 shows a section of vegetation across a river meander to an area underlain by permafrost. This is characterised by low species diversity, very low primary productivity, and secondary consumption that is carried out by species which are migratory, or which spend the long winter in a dormant condition.

Arable weeds

Relatively few plant species (out of the total flora of the Earth) have successfully colonised those ecosystems (**agro-ecosystems**: see Chapter 7) which are characterised by the regular cycle of intense disturbance produced by farmers cultivating the land to grow arable crops, such as cereals. These species, known as arable weeds, have had about 8,000 years (since the invention of agriculture by people living in what is now Iraq) to adapt to the disturbance regime produced by activities such as ploughing, and various other measures aimed directly at destroying plants which act as competitors with crop plants. Many of the plants that now commonly occur as arable weeds have colonised agricultural land from habitats, which show a high degree of disturbance from natural causes. An example is coastal backshore shingle habitats, which consist of unstable banks of stones, gravel and usually poor soils. Plants living in such conditions must possess a suitable combination of survival traits needed to allow them to survive the very real risk of being crushed to death by moving rocks. Such traits may include:

- Rapid growth to maturity, thus minimising the time taken to complete a life cycle, and hence reducing the probability that the plant will be destroyed before it reaches maturity.
- Annual life form: the plant goes through a complete life cycle each year, returning to a defensive resting stage (seeds) for the duration of the most hostile period of the year (e.g. winter storms, during which shingle movement is most likely). Perennial plants, which retain their above-ground structures from one year to the next, would be highly vulnerable to this kind of seasonal disturbance.

- Production of numerous cheap-to-build seeds that are often easily distributed by the wind or by animals. This is important in a 'shifting' habitat, where the original 'favourable' location (favourable by definition since the parent plant must have successfully survived there if it has produced seeds) may become unpredictably unfavourable, for example, if it is suddenly buried under a heap of stones. In these circumstances it is a good strategy for the plant to throw out as many seeds as possible in the hope that some will find a favourable germination site.

- Possession of **dormancy** in the seeds. Dormant seeds have the advantage of being able to survive long periods buried in the soil. They will not germinate until some external factor (e.g. cold, water, light) brings about a chemical change in the seed, which stimulates germination. Effectively this creates a seed-bank of seeds of varying age lying in the soil, a proportion of which will germinate each year if conditions are right. In an unpredictably dangerous habitat this benefits the plant population because even if all the adults are wiped out one year and no seed is returned to the soil, seed from earlier years still exists to permit the possibility of continued survival.

The characters described above are ideal for surviving in cultivated conditions, where the farmer's land management practices are deliberately aimed at destroying adult weed populations. Hence species such as groundsel (*Senecio vulgaris*), or mayweeds (*Matricaria* and *Tripleurospermum* spp.) which grow well in British coastal shingle banks, and many other naturally disturbed habitats, have also colonised British cereal fields and can be nuisance arable weeds. Of the 1,043 species listed by the European Weed Research Society (Williams 1982) as the primary agricultural weeds of Western Europe, a high proportion have their original distribution in naturally disturbed ecosystem conditions (Grime *et al.* 1988).

Over the 8,000 years or so that weeds have been co-existing with crop plants it is fair to say that the intensity of disturbance produced by farming operations has steadily increased. Neolithic farmers were limited to hand-pulling and the use of simple tools such as stone-bladed sickles for destroying the weeds in their crops. The invention of the plough, and its subsequent improvement from a simple human- or horse-drawn implement to the massive tractor-drawn steel ploughs of today, substantially increased the pressure on survival of weed populations. More recently the advent of herbicides ('weed-killers': which greatly increase the stress imposed on the plants) has resulted in the imposition of very high selection pressure on weed communities and individual weed populations.

Weeds have responded to these increasing pressures on their survival in two ways. The first is that species shifts have occurred: plant species more tolerant of the combination of stress and disturbance pressures imposed upon them by agricultural operations have tended to replace the original species. A good example is the way in which 'traditional' cereal weeds such as poppies (*Papaver rhoeas*) and corn marigold (*Chrysanthemum segetum*), which had characterised British cereal fields at least since Roman times, were largely replaced during the latter part of the twentieth century by a range of weed species which had hitherto been much less conspicuous members of the cereal-weed complex in wheat and barley fields (e.g. Fryer and Chancellor 1970; Eleftherohorinos *et al.* 1985). These plants include species such as scarlet pimpernel (*Anagallis arvensis*) and common field-speedwell (*Veronica persica*). The primary cause of this was the introduction of selective herbicides, such as 2,4-D and MCPA (which kill most broad-leaved cereal weeds but do not kill cereal plants) to weed-control programmes on a massive scale across the cereal-growing areas of the UK. Highly successful in agronomic terms, the resulting loss of the susceptible weed populations opened the way for colonisation of the niches vacated by these species by plants naturally resistant to the new herbicides, hence producing a rapid species shift (within just a few years) in the weed flora of British cereal fields.

The second way in which weed populations have responded to intensive agricultural management practices is seen in those species that possess an innate genetic trait (or traits), usually expressed in the phenotype only at very low frequencies (e.g. one plant in ten million; 1×10^{-7}), which confers resistance to a particular agronomic management procedure. Such genetic shifts, in which the weed population changes from dominance by susceptible plants to a population almost entirely made up of resistant plants, can be remarkably rapid. Good examples have occurred with the development of herbicide-resistant populations of plants in crops (such as maize in the USA) where repeated use of one herbicide (e.g. atrazine: a highly effective photosynthesis-inhibiting triazine her-bicide) occurred over a run of years. The rate at which dominance by resistant species arises in a previously susceptible population of weeds is governed not only by the start-ing proportion of the resistant gene, but also by the ecological fitness of the resistant compared to the susceptible phenotype, the selection pressure imposed by the herbicide, and the average life span of seeds in the seed-bank. For a weed population starting at a proportion of 1×10^{-7} for resistant individuals, experiencing a selection pressure which killed off none of the resistants but 90 per cent of the susceptible plants before they could set seed, and with only limited dormancy (so with a short-lived seed-bank from which susceptible plants laid down before the herbicide regime began can germinate) the time needed for the population to become effectively 100 per cent resistant to an annually repeated herbicide regime can be as short as seven years (Murphy 1983).

These examples illustrate the importance of human-produced stress and disturbance pressures in influencing the plant community composition of agro-ecosystems. Informa-tion on how plant populations respond to such pressures is also highly topical given the current controversies over the introduction of herbicide-resistant genetically modified crops that permit the use of herbicides (such as glyphosate) which produce very strong selection pressures on the weed populations associated with such crops. Introduction of such extreme stress regimes might well be expected to result in even bigger impacts on the agro-ecosystems in which they are used, either in terms of community or genetic changes in their vegetation, with potentially serious implications for the maintenance of agro-ecosystem biodiversity (Robinson and Sutherland 2002; Wilson *et al.* 2003).

The land–sea interface: marine rocky shores

The intertidal ecosystem on marine rocky shores is characterised by a seaweed zonation produced by a combination of the stress and disturbance caused by tidal action (Box 6.3) modified by exposure to wave action. The stress is associated with the period of exposure to air during low tide periods. As marine algae, seaweeds are not very good at coping with the desiccation stress produced by exposure to air. Seaweed species have different tolerances of desiccation stress: those which occur higher on the shore, relative to mean sea-level, are exposed longer to air at every tide than those which typically grow lower on the shore.

Characteristically there is a three-zone pattern, each zone being dominated by different functional groups of seaweed species. The topmost zone (*littoral fringe*) occurs around the extreme high watermark for spring tides (EHWS). On British shores this zone is dominated by salt-tolerant lichens such as *Verrucaria maura*. Below this, the eulittoral zone, centred on mean sea-level, is dominated by brown fucoid algae, such as *Fucus vesiculosus*. The lowest zone (sublittoral) occurs at and below the extreme low watermark of spring tides (ELWS) and is dominated by big laminarian kelps (e.g. *Laminaria digitata*). The varied ability of plants of the different zones to tolerate

Box 6.3

Tidal cycle

The tidal cycle is a variation in tidal amplitude (the vertical distance between the low tide and high tide marks on the shore), from neap tides (lowest amplitude) to spring tides (highest amplitude). It operates on an approximate fourteen-day cycle, produced by gravitational attraction of the sun and moon. Local conditions around the world vary the size of the amplitude range. Some examples are given below:

	max. amplitude (m)
Lake Superior (USA)	0.02
Mediterranean	1
Firth of Clyde (Scotland)	8
Bay of Fundy (Canada)	> 20

desiccation stress is shown by experimental studies which show that a fucoid species, *Pelvetia canaliculata*, which typically occurs highest in the eulittoral zone, has a survival period (in air at 20°C) of more than twenty hours. In contrast *Laminaria digitata* can survive for less than two hours in air under the same conditions.

Stress thus plays a role in determining the characteristics of the plant community of marine rocky shore ecosystems. However, wave action can create a lot of disturbance in such systems. The more exposed the site, the greater the wave disturbance. This has two effects. If the disturbance is high enough (e.g. an exposed promontory site), the combination of stress and disturbance becomes sufficiently intense to push the site into the category of uninhabitable by plants. In these circumstances the colonisation sites usually occupied by the seaweeds are instead occupied by an animal, barnacles, which have armour-plating sufficiently strong to allow them to survive the battering produced by the waves. In less strongly disturbed sites the main effect of wave disturbance is to push the zones higher up the shore, relative to tidal levels. This is because waves and spray break higher on shore, allowing the seaweeds to survive higher up than is possible on less exposed sites, such as in a sheltered bay. In addition to the effects of waves, grazing (by invertebrates such as limpets, *Patella* spp.) exerts further strong disturbance pressure on the seaweeds of the marine intertidal ecosystem.

Grazing disturbance is an extremely important pressure affecting the functioning of many ecosystems. The effects of grazing and management of vegetation for grazing in an upland ecosystem are discussed in Chapter 9.

Ecosystem fragility and resilience

Ecological change, whether natural or human induced, occurs in very complex ways. Changes rarely act consistently in one direction, or at the same rate for long periods. This means that it is difficult to predict how an ecosystem will change in the future, even when good data about existing and past conditions are available. In the first chapter of this book, we saw how much of the pioneering research work in environmental and ecological science led to the development of models of change over time. Clementsian succession, as discussed in this chapter, is a good example of such a model. The Davisian cycle, which was developed at the beginning of the twentieth century, is an example of a model of systems behaviour in the abiotic environment. Davisian theory was very influential initially but was criticised by later workers. Better measurement of ecological and environmental systems in particular cast doubt on the widespread validity of this type of theory. Furthermore, these are examples of theories that do not include human impacts as components in the system. Theories that can explain and predict the relationships between ecosystems and the changing environment must include forcing factors of human origin.

As discussed in the review of the development of the ecosystem concept in Chapter 1, progress in research in ecological science has led to criticism of the ecosystem concept. It is widely accepted that knowledge of particular ecosystems at this time is insufficiently developed to allow complete prediction of outcomes of functioning. Nor have more than a few verifiable general rules about ecosystem function yet been developed. However, H.T. Odum (1983) has proposed that there are a number of trends which can be recognised as ecosystems develop. As ecological science progresses, for example, by incorporating non-linear dynamic theory, based on the use of more sophisticated techniques such as the mathematics of chaos theory, better models of the precise functioning of ecosystems will be developed and tested by empirical research. In the interim the ecosystem provides the best framework for the investigation of the interactions between the living world and its abiotic environment. It also provides a means whereby the impact of human actions on the biosphere may be identified and analysed. Without an integrative framework, the true nature of environmental change, human impacts and the threat to the functioning of the biosphere and our life support systems, which may be real or exaggerated, cannot be understood.

One concomitant of non-linear change in ecosystems is the notion of ecological and environmental thresholds. It is now generally believed that in the majority of ecological and environmental systems, processes operate in a stepwise rather than a smoothly progressive manner over time (Phillips 1992; Nillson and Grelsson 1995). A particular set of conditions is relatively stable, or meta-stable, fluctuating but remaining within boundaries for its system parameters. Externally forced and internal change is moderated by negative feedback loops, such as density-dependent population controls or sediment budgeting. However, a big enough change will cause this meta-stability to break down, and the properties of the system to alter very rapidly, often to a profoundly different condition. One of the best ways of understanding this is by examining what happens to a spring when it is subjected to a load. If the spring behaves ideally, there is a directly proportional relationship between the load or stress, and the distortion or extension of the spring. If the load is removed, the spring will return to its original condition. If, however, the load is greater than a certain value, the spring will distort, and even after the load is removed will remain distorted. If the force is big enough, the spring may break. The point at which the spring loses its ability to recover is a threshold. Beyond that threshold the spring 'system' behaves in a different way. It is much less resilient to further change. Ecosystems behave in a similar way, showing a degree of resilience to impacts. The resilience of each ecosystem is different, and is a function of its biological communities and their functional ecology. When ecological thresholds are crossed the whole ecosystem will become unstable, and liable to rapid and catastrophic change.

Rapid ecological change, which happens as critical ecological thresholds are crossed, may be a result of natural processes. The science of natural environmental change is considered fully by Mannion (1999) in the book *Natural Environmental Change* in this series. Examples include damage to biological populations by disease or parasitic infestation, or the effects of a landslide or an extreme climatic event such as a storm. In most cases not all individuals are affected, but great epidemics may devastate whole populations. Trees, as the case of Dutch elm disease shows, despite their size and persistence, may be as vulnerable as smaller organisms. In cases where **keystone species** are affected, the whole community is likely to experience change in both species composition and numbers. Coastal erosion provides a good example of environmental change in the physical environment that acts in this way. Over long periods of time, change at the coast is generally subdued, fluctuating around a particular set of conditions. Beaches will react to seasonal weather conditions through cyclical change in profile and sediment characteristics. However, a single severe storm may cause permanent change to the whole system by breaching dunes, removing sand and modifying the balance between sediment load and transport energy. The above circumstances may be triggered or accelerated by human impacts. Generally, human impacts cause damaging change to ecosystems more frequently than do normal processes. Furthermore, research indicates that the rate *and* intensity of human impacts is accelerating. It is widely recognised that reducing human impacts, which are damaging to ecosystems, is one of the greatest challenges humankind has ever faced. These challenges are considered more fully in Chapters 9 to 11.

The spatial patterns of vegetation and ecosystems

The early scientific study of vegetation as a whole, rather than individual species of plants, was largely through **phytosociology**. This was based on the premise that there were recognisable assemblages of vegetation, defined by species composition, which would be found in particular locations. Two broad schools of approach developed, both in Europe: the Zurich-Montpellier and the Uppsala. The Zurich-Montpellier school advocated selection of 'typical' sites using subjectively identified recurrent vegetation assemblages for their definition. The Uppsala school had a rather more quantitative approach using **quadrat** vegetation samples to describe the assemblage. These are generally termed 'associations'. This approach has been criticised in that essentially it is subjective. This criticism is valid. Whittaker (1956) showed that there were no distinctive plant assemblages in forests in the southern USA. Loucks (1962) demonstrated that forest vegetation was not composed of discrete units, but was related to environmental gradients and that the phytosociology of each part of forest was, in detail, unique. These and many other pieces of research showed empirically and objectively that vegetation varies continuously in space – it was a **continuum** – and that associations in the earlier sense of a recognisable repeated spatial unit did not exist. Much research work relating to the continuum was done in true natural vegetation, particularly in tropical, subtropical and warm temperate forests. Colinvaux has stated, with much validity, that if the ecosystem was the major advance, which came from phytosociology, the notion of discrete vegetation communities with discrete spatial boundaries was its greatest shortcoming (Colinvaux 1993: 410).

While nowadays the defining paradigm on the nature of vegetation is that of the continuum, spatial patterns of vegetation remain an intriguing and important topic. Much current ecological work on plant community – environment relationships routinely use both concepts (gradient analysis and plant associations) together as tools for analysing vegetation data. The use, as complementary approaches of programs such as CANOCO (which provides ordination tools, such as canonical correspondence analysis) and TWINSPAN (two-way indicator species analysis: a classification program which establishes units of vegetation and characterises them in terms of species which indicate each unit) for this purpose is now widespread in plant ecological studies (e.g. Murphy *et al.* 2003). Much ecological work depends on good and accessible spatial information on biota. Maps are important, because they are comprehensible to a wide range of users. Spatial patterns may be described by means of ordination techniques, but not easily in a Euclidean way which non-specialists – decision-makers perhaps! – can understand. Transects and maps which use some form of classification of vegetation remain important tools. For the latter, classification of vegetation types into discrete units is useful. This does not imply an inherent natural structure of associations, but may simply be an appropriate way of handling and communicating data. One recent such method is the British National Vegetation Classification (NVC: Rodwell 1991a, 1991b, 1992, 1993, 2000), developed in the 1980s by classification of vegetation data using TWINSPAN, and now widely used to characterise all types of vegetation found in the British Isles. The development of Geographical Information Systems (GIS) which are powerful tools for analysis of spatially referenced data, and the Global Positioning System which makes acquiring accurate positions very rapid and simple reinforce the value of this kind of approach.

There is one final factor about the nature of vegetation which needs to be considered. This is the role of humans in causing actual patterns of vegetation. We talk about natural vegetation, and by implication, ecosystems. Humans are seen as an extraneous factor. Without getting into an extended philosophical debate about this, two facts need to be considered. First, humans have been modifying ecosystems for a long time. The hominid use of fire pre-dates *Homo sapiens*, and fire-modified vegetation has been a reality for at least several tens of thousands of years. Second, the effects of human actions are felt to a greater or lesser extent throughout the biosphere. This may or may not be a bad thing, but it is a reality. Thus human actions need to be considered in the analysis of contemporary ecosystems. The effects of human activities in many cases apply in more or less discrete areas in which the actions have taken place. Examples of this include burning, grazing and clearance of forests. The resulting vegetation patterns, even if largely now under the influence of purely 'natural' process, inevitably are spatially discrete to some extent. It is noteworthy that the early phytosociological approach to vegetation developed in Europe, the vegetation of which has been influenced by humans, often profoundly, for millennia. In comparison the pioneering work using ordination techniques was carried out in forests scarcely modified by human activity.

In the next chapter (Chapter 7) we examine the spatial pattern of vegetation on a global scale. Biomes are the outcome of interaction of many environmental factors, though at a global scale climate tends to be the dominant factor. Biomes are not characterised by particular species but by life forms which in turn are adaptations to environmental conditions. These are thus functional groups. But again we must remember that throughout the world ecosystems have been modified by human actions. The nature of these modifications, their effects on ecosystems, have significance for humans and all life on Earth.

Summary

- The subject of this chapter, disturbance to ecosystems, has both natural and human components. Natural disturbance is a feature of the early stages of some types of vegetation succession, as well as being found in parts of more mature systems.
- Plants that have adapted to disturbance are able to dominate communities in a wide range of physical environments, and are well placed to survive in situations in which human actions have caused disturbance.
- Examples of colonisation on lifeless surfaces and at the land–sea interface show how ecosystems respond to disturbance of both natural and human origin.
- The nature of system dynamics and resilience to change are important elements in ecosystem functioning.
- The ways in which the study of change in vegetation in both time and space have influenced thinking about the nature of vegetation systems is discussed.

Discussion questions

1 Disturbance is a relatively constant factor in some ecosystems, but in others disturbance generally decreases over time. Give examples of each, and analyse the differences that may be detected in your examples, in the development of functional types of vegetation and soil conditions.

2 Change in sea-level causes disturbance to ecosystems at the interface between land and sea. As raised sea-levels are an almost inevitable consequence of global climatic change caused by human actions, there is much concern about this. Give two examples of the effects of such disturbance on a coastal ecosystem (either aquatic or terrestrial), and comment on whether or not this will result in damage to these ecosystems.

3 Can you think of any instances in which human disturbance is a beneficial factor for ecosystem function? Remember that disturbance may be a deliberate action in environmental management.

Further Reading

See also

Organism–environment interactions, Chapter 2
High stress environments, Chapter 5
Biomes and functional ecology, Chapter 7
Human impacts on ecosystems – impacts on trophic structure, Chapter 9
Large-scale impacts on ecosystems, Chapter 10

Further reading in the Routledge Introduction to Environment Series

Biodiversity and Conservation
Environmental Biology
Natural Environmental Change

General further reading

Biodiversity and Ecosystem Functioning. Ed. M. Loreau, S. Naeem and P. Inchausti. 2002. Oxford University Press, Oxford.

A challenging but extremely informative review of current knowledge in this important area.

Ecology of Salt Marshes and Sand Dunes. D.S. Ranwell. 1972. Chapman and Hall, London.

A classic, comprehensive analysis of these ecosystems at the land–sea interfaces.

Plant Strategies, Vegetation Processes and Ecosystem Properties (2nd edn). J.P. Grime. 2001. Wiley, Chichester.

Chapter 8 (Succession) gives an elegant review of this topic from a functional ecology perspective.

Studies in Plant Survival. R.M.M. Crawford. 1989. Blackwell, Oxford.

This contains a series of case studies relating to plant life in disturbed environments.

7 Life in a crowd: productive and intermediate ecosystems

The harsh environmental conditions with which plants, animals and micro-organisms have to cope in highly stressed or disturbed ecosystems are not encountered by the great majority of species. The highest biodiversity of species occurs in the more kindly conditions of intermediate ecosystems, often with a mosaic patchwork of differing combinations of conditions, supporting a variety of species. In the best conditions of all for growth (the most productive ecosystems), biodiversity drops again, because the most competitive species tend to oust their neighbours from such ecosystems. This chapter covers:

- **Defining competition**
- **High production ecosystems**
- **Relationships between competition and productivity**
- **Intermediate ecosystems**

Defining competition

Competition between organisms within the habitats making up an ecosystem has been defined in many ways, but Keddy (1989) has provided a succinct and clear definition: see box.

> **Definition**
>
> Competition: the negative effects which one organism has upon another by consuming, or controlling access to, a resource that is limited in availability (Keddy 1989: 2).

Competition may technically occur between any pair of organisms, whether they are from populations of the same species (intraspecific competition) or drawn from populations of different species (interspecific competition). However, in practice, competition occurs only when two populations compete for a *resource in limited supply* which is necessary for the survival of each. In these circumstances there is a tendency for the more competitive population to exclude the less successful one. Early experimental work on yeast (Gause 1932) and beetles (*Tribolium*: Park 1954) in limited-resource experimental systems suggested that competitive exclusion is a general principle in ecology. Pairs of very similar species (in terms of size and environmental requirements, i.e. having closely similar niches) find it difficult to coexist in the same ecosystem because competitive pressures between them are too strong. Ecological differentiation (Hardin 1960) appears to be necessary for species to coexist in crowded, competitive ecosystems. In practice, coexistence seems to be what happens in real (i.e. not artificial experimental) ecosystems. Competitive exclusion is rarely seen to occur to the bitter end, and ecologists have devoted enormous effort, and much imagination, in trying to develop models which can successfully explain the coexistence of species. These are discussed further by Keddy (1989).

With a few exceptions, major (i.e. broadly distributed, dominant, successful) species must be good competitors. The exceptions are those which have successfully colonised extensive stressed or disturbed environments, where interspecific competition pressures are low and possession of genetic traits for tolerance of stress- or disturbance-related pressures on survival are, instead, at a premium (see Chapters 5 and 6). Good examples are those mosses (e.g. woolly-fringe moss, *Rhacomitrium lanuginosum*) which have adapted to the cold conditions of high-latitude upland and tundra areas, and which are common and widely distributed plants in these circumpolar and alpine cold-stressed habitats.

High production ecosystems

Where environmental circumstances are favourable for life, particularly where temperature conditions provide good, all-year-round conditions for photosynthesis, where water supplies are abundant and where general nutrient availability is good, then ecosystems tend to support species which are capable of achieving high rates of production (Box 7.1). Organisms in such ecosystems live in crowded conditions; the main threats to their survival tend to be from biotic, rather than abiotic, pressures. Obtaining the resources needed, even when these are in abundant supply, may be rendered difficult because of competition from more efficient organisms for the same set of resources.

Box 7.1

High competition ecosystems

Stress and disturbance *low* (e.g. warm temperatures, high light intensity, plenty of water – tropical rainforest)
↓
good growth conditions: productive ecosystem
↓
main problem faced by plants: *other plants*
(competing for the same set of resources needed for growth)
↓
high rates of resource depletion

- Plants face life in a crowd
- *Competitive strategy* needed: right combination of traits to allow plant to forage effectively for the resources it needs, in the face of this strong competition
- Successful C-strategist may rapidly grow tall, for example, with a dense leaf canopy (excludes light from potential competitors) and well-ramified roots

Competition and productivity

There are strong links between competition and biological productivity. This is partly a function of the opportunities for growth provided by the physical environment, and partly related to the response of biological producers and consumers to these opportunities through the processes of competition and predation. In plant communities, researchers have tried to identify the combinations of traits which help a plant population to be competitive (e.g. Gaudet and Keddy 1988). Once such traits have been identified they may be used as predictors of plant success in different ecosystem conditions. For example, in riverine wetland ecosystems of Western Europe, Hills *et al.* (1994) found that certain plant traits, such as height and leaf size, could be measured in field populations of the plants, and used to identify functional vegetation types (see Chapter 2) which showed differing competitive and stress-tolerance abilities.

The humpback model

In terms of ecosystem functioning it is worth noting that one major prediction of competition theory is that the biodiversity support function of ecosystems appears to be greatest at *intermediate* intensities of stress and disturbance, where a large number of niches are open to colonisation. The most productive ecosystems do not necessarily support the highest diversity of species. This relationship follows a typical *humpback* shape (Grime 1979; Ali *et al.* 2000).

Take, for example, the case of submerged freshwater plants ('macrophytes') growing in Swiss lakes (Lachavanne 1985). Only a few stress-tolerant species (mainly isoetids: see Chapter 2) occur in nutrient-stressed ultraoligotrophic lakes. The plant diversity increases steadily as nutrient status increases, but only up to a point (generally around mesotrophic conditions, i.e. moderate–high availability of nutrients). Beyond this point the macrophyte diversity starts to collapse as the lakes move into eutrophic, then hypertrophic conditions. In the most nutrient-rich, highly productive lakes (hypertrophic conditions), only a handful of macrophyte species occur, or even none at all. Here the productivity emphasis shifts to massive blooms of phytoplankton concentrated in the surface layers of the water, which outcompete the submerged macrophytes for light. The green-pea soup conditions which they create provide very hostile (i.e. very low) energy conditions for submerged macrophytes trying to grow in the water. The dense crowd of phytoplankton (which may reach concentrations of a million or more cells per millilitre of water: see Box 7.2) absorbs much of the down-welling light entering the water, severely reducing both the quantity and quality of available light energy for plants growing below or within the bloom (see Chapters 3 and 5). The compensation depth in such lakes may be very close to the surface: 1 m or less. This often has the effect of reducing the area of the lake where macrophytes can grow to a narrow band of shallow water closest to the shore, further reducing the potential number of habitats available for macrophyte species to occupy, and hence further reducing the diversity support function of the lake ecosystem.

Similar relationships for diversity vs. productivity have been observed in many other ecosystems, both aquatic and terrestrial, and although they can be partly explained by density-dependent factors, biodiversity does appear to be quite closely predicted by such 'humpback' models. Competitive exclusion (by competition for available resources: often, but not exclusively, light in the case of plants but other resources for other organisms) probably plays a role in reducing the diversity of species occurring in the most productive habitats.

Box 7.2

Phytoplankton

The pelagic phytoplankton communities which form algal blooms are phylogenetically diverse and have highly complex structure and dynamics, both in time and space. The diversity of organisms may be high, especially in waters experiencing intermediate frequencies of disturbance (such as mixing by currents: Padisak 1993). Seasonal factors (e.g. the annual occurrence of thermal stratification in temperate lakes) can produce cyclical changes in the predominance of different groups, which are a function of the changing temperature, physical structure and nutrient availability of the lake water body during the year.

An excellent summary of the ecology of phytoplankton in freshwater ecosystems is given by Moss (1988). He gives an interesting analogy which helps us understand the scale of the universe in which the phytoplankton live (Figure 7.1). If the smallest phytoplanktonic unicells (so-called picoplankton: no more than 1–5 μm in diameter) are taken to be the size of children's marbles, then the largest phytoplankton colonies (such as *Volvox*, which forms balls of aggregated cells up to 500 μm across, just visible to the naked eye) are the size of an elephant. The largest of the herbivorous zooplankton which graze the algal blooms (mainly Crustacea, such as the water flea, *Daphnia*, up to 3 mm long) would be house-sized in our analogy, while the fish which swim through the blooms would be ocean liners! In this scaled-up model the crowded conditions occurring

Figure 7.1 Relative sizes of bacteria, phytoplankton and zooplankton: (A) a bacterium; (B) *Cryptomonas*, a small phytoplankter; (C) *Scenedesmus*, a moderately large phytoplankter; *Keratella* (a rotifer: a fairly small zooplankter; (D) the head and eye (e) of *Daphnia*, a large zooplankter (the head is about a quarter of its total body size)

in water containing a dense bloom of the smaller species of algae could be visualised as a glass-sided squash court completely filled with a crowd of green balloons tethered to float a metre apart from each other. Seeing through the squash court would not be easy. By analogy the submerged macrophytic plants occurring beneath the bloom also 'see' little of the light down-welling from the surface.

Like their larger relatives (macrophytes and land plants) the microscopic phyto-plankton can be classified on size and morphological traits using CSR terminology (Reynolds 1996). Examples of stress-tolerant (S-strategist) phytoplankton include the larger green algal colonial forms, like *Volvox* and nitrogen-fixing filamentous Cyanobacteria such as *Anabaena*. Smaller (i.e. with a high surface:volume ratio) uni-cellular green algae such as *Chlorella* are competitive (C) species. Disturbance-tolerant R-strategists include many small, fast-reproducing diatoms such as *Melosira* and *Asterionella*. Some of these also form filamentous or (in the case of *Asterionella*) star-shaped colonies of just a few cells.

The pelagic plants and bacteria of the phytoplankton are influenced by the same basic environmental pressures of stress, disturbance and competition as are larger plants in terrestrial vegetation. The community dynamics which govern the success or otherwise of the different phytoplankton strategists in aquatic ecosystems are closely analogous to those occurring on land (or in macrophyte vegetation in water). But there is one import-ant difference: all the community processes 'happen absolutely much more quickly; only once recognised the plants of the pelagic are the perfect scale models of vegetation processes' (Reynolds 1996).

Examples of productive ecosystems

Tropical rainforest ecosystems

Tropical rainforests provide good examples of crowded, productive conditions for plant growth. Their productivity pushes towards the extreme at which high biodiversity can occur (see Table 3.3). Ecosystems with productivity higher than that of the rainforests (such as water hyacinth-covered lakes) tend to show markedly reduced biodiversity. The tropical rainforest biome is confined to equatorial regions of the world, mainly occupying the hot, humid, low-lying basins of major river systems such as the Congo and Niger in the African rainforest, and the Amazon and Essequibo Rivers in the American rainforest (See Chapter 8). The third major area of rainforest (the Indo-Malayan region) is less cohesive, being scattered across parts of continental Asia (mainly coastal areas, such as Vietnam and the Malayan peninsula), the islands of Indonesia, and as far south as southern Queensland in Australia (where a quirk of local conditions allows a tropical rainforest ecosystem to occur as far from the equator as 27°S).

In the classic rainforest conditions found, for example, in Borneo (hot, wet, plenty of incoming light, although with relatively poor-nutrient soils), the key to success for plant species is to grow as tall as possible to outcompete neighbouring plants. As a result, rainforest tree communities tend to be dominated by species such as the very tall and very large dipterocarp trees, which form the top canopy of photosynthetic tissue in such plant communities – i.e. they have first call on the incoming PAR. In all, such forests may have up to seven discernible strata or stories of above-ground plants, with the lower layers being composed of species adapted to low levels of light availability.

Overall the biodiversity of such forest ecosystems is famously high compared with many other ecosystems. While a sample 200 m² plot of seasonal riparian rainforest in southern Brazil had more than twenty species present, a larger plot (60 × 7.5 m) of pine forest in the much higher stress conditions of the Scottish Highlands might well have only one or two tree species present: Scotch pine, *Pinus sylvestris*, and maybe rowan, *Sorbus aucuparia*). The forest floor component of the rainforest ecosystem, beneath the multilayered tree canopy, is rather akin to conditions below a dense algal bloom in a lake. Here there is a low-energy habitat for plant survival, and also the soils tend to be rather deficient in nutrients because so much of the available nutrient supply is locked up in the trees' biomass. That a much higher diversity of plants per unit area can be achieved in a rainforest is seen in natural clearings, where a tree has fallen and opened up a patch of ground to the sky. Here there occurs a profusion of species, often including orchids and other low-growing species. Competitive exclusion, by the trees, appears to be just as important here as in the Swiss lakes example, in reducing the expressed diversity of plants under conditions of high competition for light.

Wetland ecosystems

Wetlands are often rather productive systems, where competition may be quite intense. Their productivity varies, and as in the case of tropical rainforests, those wetland systems with the highest productivity tend to have reduced biodiversity (e.g. near monospecific *Phragmites* reedswamp, which can have extremely high productivity: see Table 3.3). They can achieve such high productivity because water is often not limiting to growth (though there are variations in water availability within a wetland ecosystem), and nutrient availability may be high.

In European riverine wetland ecosystems there is evidence that functional groups of plants exist which are differentially adapted to the different intensities of stress vs. competitive conditions which occur within and between such systems (Hills and Murphy 1996). For example, in many Spanish wetlands stress is quite high: the wetlands tend to dry out in summer (a problem made worse by the declining groundwater levels in many parts of Spain due to over-extraction for irrigation purposes) and the remaining water may become quite saline (due to evaporation, which concentrates the salts present in the remaining water). In these wetlands a stress-tolerant group of wetland plant species, of rather low diversity, tends to occur. Elsewhere, for example, in the wetlands (or 'callows' as they are locally known) bordering the River Shannon in central Ireland, vegetation shows an overall much lower incidence of expression of stress-tolerance traits, and the main variation in plant functional group type is strongly related to variations in topography (which affects the probability of inundation by floodwaters) and groundwater levels (influencing the probability of exposure to summer drought conditions) across the wetland ecosystem (Hills *et al*. 1994; Hooijer 1996). The mosaic of vegetation which results from this variation in conditions within the Shannon callows gives rise to a high diversity of plant species – in line with the predictions of the humpback model.

Work done in wetland habitats fringing a Canadian lake (Wilson and Keddy 1986) provides experimental evidence to support the idea that traits for competitiveness trade off against those conferring tolerance of stress or disturbance in wetland plants. Axe Lake in Ontario has a range of shoreline types, forming a gradient of environmental conditions from high-disturbance (due to wave action), low-nutrient habitats on exposed, gravelly shores of promontories, to the low-disturbance, high-nutrient conditions of sheltered bays where silts accumulate to give better soil conditions. By setting up an experimental series of pairwise combinations of seven wetland species, at different locations around the lake shores, Wilson and Keddy were able to show that the plants differed substantially in their competitive ability (measured as relative biomass increase when grown together) at different points along the environmental gradient. These differences were clearly related to the actual habitats occupied by the different species around the lake shores. Thus, for example, *Dulichium arundinaceum* (found in the sheltered, most productive sites) had the highest competitive ability of the seven species compared. In contrast, amphibious isoetids (see Chapter 2) like *Eriocaulon aquaticum*, which occupy the less productive, more exposed habitats around Axe Lake, were poor competitors.

Subsidised agro-ecosystems

Agro-ecosystems vary considerably in the amount of subsidy they receive from agricultural activities in terms of energy or material inputs (see also Chapter 6). At one extreme are natural rangeland ecosystems, where management is minimal or non-existent, and the natural vegetation is being utilised for animal production (e.g. cattle): such ecosystems are usually somewhat stressed and nearly always disturbed by grazing pressure. Dry prairie grassland ecosystems are a good example of these low-productivity agro-ecosystems. With increasing subsidy, productivity rises, and by definition so does the intensity of competition experienced by the plants occupying the ecosystem. There is a gradient of increasing production running from:

- a managed grassland, receiving low quantities of N-P-K fertiliser to subsidise the growth of grasses for the grazing animals, through
- an arable system, such as a wheat crop, receiving much higher fertiliser subsidies (and producing perhaps 120–200 tonnes ha^{-1} of vegetation per year, with high competition from the crop plants against any other plant species present) to
- a high-subsidy agro-ecosystem such as a horticultural glasshouse crop of tomatoes, where is there not only a heavy input of fertilisers, but also often an energy subsidy in the form of artificial lighting.

In general the more heavily subsidised the agro-ecosystem, the more productive it will be, and the greater the intensity of competition. The farmer or grower is interested only in maximising the productivity of the crop plant species, and there are likely to be additional discouragements to the growth of other species (weeds), for example, herbicide spraying. Herbicides are a form of indirect subsidy to the crop plants, designed to reduce the competition for available resource from weeds, and thereby increase the production achievable from the cropland area. Herbicides actually work by placing the weeds under severe toxic stress, while leaving the crop plants (more or less) unaffected, a good example of a *selective* stress pressure acting on the plant community. In these non-natural ecosystems intra-specific competition (between the plants of the crop population) may act as the limit to production.

The humpback model would predict an inverse relationship between ecosystem biodiversity and intensity of stress or disturbance in such agro-ecosystems. Certainly for the vegetation this appears to be true. In Scotland, it has been shown that the diversity of the plant community – measured either as species richness (the number of species present) or using a diversity index (such as Shannon's index, which takes account of the relative abundance of species present) – on different types of agricultural land is related to the intensity of management affecting the vegetation (Abernethy *et al.* 1996; Wilson *et al.* 2003). Higher disturbance is associated with more intensive management. In the low-intensity management conditions of upland sheep-grazed grassland in the Scottish

Highlands the diversity is quite low. Plant diversity rises sharply as we drop down to the semi-natural vegetation of long-term managed grassland in the glens, then starts to decline as management intensity increases still further into the heavily managed short-term grass leys of lowland cattle-rearing areas. In the most intensively managed agro-ecosystems (intensive arable croplands growing barley or oilseed rape) plant diversity is extremely low (the crop itself plus a handful of weed species), although it is increased if the non-arable parts of the land (such as field boundaries, hedgerows) are taken into account. Such areas act as an important refuge for plant species in otherwise heavily subsidised ecosystems.

Intermediate ecosystems

In this chapter and Chapters 5 and 6 we have described some examples of extreme conditions in ecosystems. In these situations sets of organisms strongly adapted to stressed or disturbed conditions, or species which are both highly competitive and highly productive are the successful occupants of the ecosystem, depending on the particular set of extreme conditions prevailing.

Simply because these *are* extreme conditions, such 'single-pressure' ecosystems are relatively unusual within the biosphere as a whole. Some ecosystems which tend towards the extremes certainly do occupy extensive areas (up to biome scale): the cold-stressed conditions of the tundra offer a good example. However, even here disturbance pressures exist (such as the habitat disturbance produced by cryoturbation, and grazing by herbivores such as lemmings and reindeer: see Chapter 6). The highly productive conditions of wheat-growing arable lands, stretching over large areas of Canada and eastern England (to name just two productive cereal-growing areas) provide an example of an extensive competitive environment for plant growth. But again, there is also an important element of disturbance, in this case inherent in agricultural management practices such as ploughing.

Most ecosystems provide life support conditions for their biota which are intermediate between the extremes. In these ecosystems most organisms experience a degree of crowding which is closely related to the productivity of the ecosystem (as outlined earlier in this chapter), but which is modified by the intensity and pattern of stress and disturbance conditions prevailing across the ecosystem in time and space. These modifications all tend to increase the number of niches available for colonisation by species within the ecosystem as a whole, thereby increasing the biodiversity support function of the ecosystem.

How do intermediate ecosystems provide increased niche availability?

There are three principal ways in which modification of environmental conditions may lead to an increase in niche availability in intermediate ecosystems.

First, conditions within the ecosystem as a whole may be *intermediate*, in the sense that moderate stress and/or moderate disturbance may be produced by one or several causes. So organisms with the appropriate intermediate survival strategies to cope with such pressures will tend to predominate. Among plant species, we would expect SR strategists to be successful in such circumstances. An example is the vegetation found on the thin, rather low-productivity soils around parts of the Mediterranean (Grime

1979). Here summer drought is quite intense (though nowhere near as bad as in hot deserts) and in the herbaceous vegetation growth is more or less confined to the moist, cooler conditions of winter. These plants are geophytes: they survive the summer high-stress period as underground storage organs, such as bulbs in the case of spring squill (*Scilla verna*) and rhizomes in cowslip (*Primula veris*). The combination of moderate stress, plus moderate disturbance (mainly from fire and grazing – especially by goats and sheep) means that this geophyte SR strategy is common in the terrestrial ecosystem type which borders much of the Mediterranean: for example, in Greece.

Second, conditions may show *spatial* variation across the ecosystem. There may be a mosaic of differing intensities of stress and disturbance, in differing combinations, and perhaps from differing sources in the individual habitats ('patches') comprising the ecosystem (see Case Study 5). This example illustrates how much variation in individual habitat conditions may occur in a given ecosystem. In this case a spatial mosaic of habitat components makes up quite a varied set of combinations of stress, disturbance and competitive pressures within the mountain ecosystem, which is reflected in a varied set of plant community types, in turn supporting a wide range of animal communities (e.g. heather and grouse moorland community) across the ecosystem as a whole.

Third, conditions may vary *temporally* across the ecosystem, producing a changing balance of stress/disturbance/competitive conditions over a period of time. If seasonally predictable, such changing conditions may produce alternating dominance of different sets of organisms at different times of year within the ecosystem (as, for example, in the phytoplankton of temperate thermally stratifying lakes). Migration is a common feature of such ecosystems. During the more productive periods, more competitive species (usually, but not exclusively, animals simply by virtue of their higher motility) arrive to take advantage of the high production. During less productive periods (e.g. winter in high latitudes; summer in lower latitude, dry and hot ecosystems) the migratory organisms absent themselves from the ecosystem in favour of better conditions elsewhere. Perhaps the most extreme example of an organism with this survival strategy (aimed at maintaining itself in the conditions ideal for its own requirements) is the Arctic tern (*Sterna paradisaea*), which twice a year migrates almost from pole to pole in search of the fleeting Arctic (and Antarctic) summers.

During the more productive periods of the year, in these seasonally changing ecosystems, competition may be quite intense for the organisms which periodically occupy them. For example, in the savannah ecosystems of East and Southern Africa the arrival of the huge migratory herds of herbivores such as zebra, antelope and wildebeest coincides with the onset of good vegetation growth after the rainy season. Potentially, competition between these herbivore species is intense. But the ecosystem has been in existence long enough to allow both the plant–animal interactions and the animal–animal interactions (of both the species competing for the available grazing, and their predators) to sort the organisms concerned into functional groups. The existence of these minimises the intensity of competition (and indeed predation) to produce only acceptable damage (from the point of view of the plants and prey animals) to the food resource which they represent.

Thus elephant do not normally compete directly with wildebeest for food: their niches are sufficiently separated to minimise such interaction problems. However, if stress or disturbance in the normally productive grassland system of the savannah is increased (as happened during the early 1990s in Southern Africa, for example, where elephant have increased substantially in numbers through misguided bans on culling, in part owing to pressure from well-meaning but ill-informed conservation interests), then the delicate balance between the competing herbivores collapses. The resulting increase in disturbance

Case study 5

In a mountain ecosystem characterised by broadly arctic-alpine conditions, such as the Cairngorm Mountains of Scotland, moderate-to-high stress conditions prevail across much of the high plateau area (see Chapter 2). Here S strategists dominate the vegetation. On the steeper slopes where unstable scree occurs, the intensity of disturbance is higher than on the mountain plateau. At higher altitudes in the scree habitats, the combination of cold, harsh conditions, plus the generally steeper angle of the slopes, combine to push the habitat into the uninhabitable category for plants, as discussed in Chapter 2. Lower down, however, the intensity of disturbance is reduced because the shallower gradient reduces the instability of the scree blocks (so reducing the chance of plant destruction by crushing under moving rocks). At the same time the lower altitude produces less hostile stress condition. Hence SR strategists can colonise into an intermediate habitat type within the totality of the mountain ecosystem. A good example of a successful plant in this intermediate ecosystem compartment is holly fern (*Polystichum lonchitis*: Figure 7.2) which occupies the damper crevices between boulders in the more stable scree slopes, but can cope with neither the high disturbance of very mobile screes, nor with the intense physiological drought stress produced by wind on the open mountain plateau.

In the lowermost part of the scree slopes, especially in areas of nutrient-rich rock (such as the schistose rocks of Ben Lawers, south of the main Cairngorm massif) stress problems are reduced still further because of the relatively high availability of nutrients leaching from the rather soft rock. Here plants more typical of productive-disturbed habitats, such as arable weeds, more commonly found

Figure 7.2 Holly fern (*Polystichum lonchitis*)

in the agricultural fields of the glens between the mountains, may successfully colonise. Such CR strategists occurring on Ben Lawers include annual meadow grass (*Poa annua*), common mouse-ear chickweed (*Cerastium vulgatum*) and coltsfoot (*Tussilago farfara*). More typically, these little herbs, plus a range of herbage grasses, occupy perhaps the most intermediate of all habitats within the mountain ecosystem. This comprises the valley bottoms and gentler lower-grazed hill slopes. Production here is higher than on the mountains proper, but still low compared with true lowland systems. Disturbance remains moderately high, produced by sheep and deer grazing. In consequence plants with strategies in the CR and CSR categories will tend to predominate. Bent grass, *Agrostis capillaris* (CSR: Grime *et al.* 1988), is a classic intermediate strategist, and is a dominant species in the vegetation of this type of habitat in many of the glens of the Scottish Highlands.

causes serious damage to the producer component of the ecosystem – the vegetation – with major knock-on effects for the support functioning of the savannah ecosystem across large areas of Africa.

Predicting the support functioning of ecosystems

It is apparent from the examples discussed above that most ecosystems are neither simple nor easily categorised. It is more accurate to say that most ecosystems are variable, dynamic entities in terms of the survival pressures they exert on the biota they support. The patch dynamics of these ecosystems may produce a rapidly changing and spatially variable set of pressures influencing the survival of the biota which they support.

The beauty of the CSR model of biota–environment relationships is that it provides a coherent framework to allow both simple description and more complex modelling of these pressures. The CSR model lets us examine the way in which these pressures change across time and space, and predict the responses of functional groups of organisms which experience these challenges to their survival. The CSR approach is applicable both to simple single-pressure systems (e.g. parts of the Antarctic) and also to the more complex multi-pressure ecosystems which occupy most of the planet's biosphere.

Understanding the ways in which organisms respond, in functional terms, to the balance of stress and disturbance pressures influencing their survival can give us an important key to understanding the support functioning (e.g. biodiversity support) of ecosystems. The methods of functional ecology, which use analysis of trait sets held in common by functional groups of plants, animals or micro-organisms, as the basis for this understanding, provide an important new insight into how ecosystems function. After all, the organisms which live in an ecosystem by definition must have an integrated response to all the challenges which that ecosystem offers to their continued survival. Otherwise those organisms would simply not be there. If we can quantify that response using suitable functional measures, then we have a chance of being able to develop working models of ecosystem functioning – both in operational and support terms. Such biota-based models can be extremely powerful tools for helping us to predict how ecosystems may respond to changes produced by human or natural causes in the future.

In the final chapters of this book we look at some of the implications of such changes for the functioning of ecosystems.

Summary

- This chapter discusses the characteristics of those ecosystems (and parts of ecosystems) which experience environmental conditions more favourable to survival than the stressed and disturbed ecosystem conditions described in Chapters 5 and 6.
- In these more productive conditions life becomes more crowded, and biodiversity increases follow a humpback relationship with increasing productivity (the most productive ecosystems, with very low stress or disturbance to limit growth, tend to have lower biodiversity than more intermediate ones).
- The essential ability needed for life here is to forage effectively for resources in the face of competition for the same resources from the crowd of neighbouring organisms all trying to live in the same neighbourhood.
- Examples are given of competitive conditions in tropical forest, wetland and agricultural ecosystems, and in ecosystems showing a variety of intermediate environmental conditions.

Discussion questions

1   Is competition between populations of organisms underrated by ecologists as a regulatory control acting on biota within ecosystems?
2   What are the critically important features of a competitive ecosystem?
3   Is global warming likely to lead to an increase in competitiveness in ecosystems generally?

Further Reading

See also

Definitions and examples of competitors and intermediate strategy organisms, Chapter 2
Stressed ecosystems, Chapter 5
Disturbed ecosystems, Chapter 6

Further reading in Routledge Introductions to Environment Series

Biodiversity and Conservation
Environmental Biology
Natural Environmental Change
Oceanic Systems
Wetland Environments

General further reading

Competition. P.A. Keddy. 1989. Chapman and Hall, London.

A well-written and succinct account of current knowledge about how competitive interactions occur in sets of organisms.

8 Biomes: world ecosystem types

The surface of the Earth is occupied by distinctive plant and animal communities – biomes – occupying large areas, and with broad functional ecological characteristics which address the challenges and opportunities of the environments in which they are set. The pattern of biomes is essentially controlled by primary production, which provides the energy base for the biological communities that make up each biome. There are wide variations in productivity, and thus great variation in the types of plants, which form the dominant biota in a biome. Primary biological productivity is controlled by a series of abiotic environmental factors. In terrestrial environments, heat and water availability are the most important at a global scale. Thus the world pattern of terrestrial biomes is closely related to global climates. Human influences are becoming an increasingly important element affecting the nature of biomes. This chapter covers:

- **What are biomes?**
- **Biomes and the factors which control primary production**
- **The global pattern of biomes**
- **Variations within biomes and the human factor in biomes**

What are biomes?

Biomes as defined in the glossary are regional-scale assemblages of ecosystems. To some extent there is congruence between the two concepts. However, biomes are the largest scale of unit, and as is discussed in this section the role of climate in defining and controlling terrestrial biomes is regarded as fundamental. In this chapter the reasons why these patterns develop are examined, and terrestrial biomes are analysed in detail. Although there is no detailed treatment of aquatic biomes, this should not be taken to mean that these are unimportant. Quite the reverse is true. Oceanic and other aquatic ecosystems are responsible for about half of all photosynthetic activity, and probably most chemosynthetic production. Furthermore, there are complex and interactive links between marine ecosystems and energy and material flows, as has already been discussed in Chapters 3 (Energy flow and energetics) and 4 (Material cycles in ecosystems). Links between climate and oceanic circulation are also profound. What humans do affects the oceans, and their ecosystems, almost as much as terrestrial environments and ecosystems. However, our knowledge of oceanic systems is much poorer than what is known about land ecosystems. This lack of knowledge certainly needs to be addressed if humans are to be able to understand the processes controlling the natural systems of the planet on which our lives depend.

Terrestrial biomes are usually defined climatically (e.g. desert biomes), or in terms of their dominant vegetation form (e.g. rainforest biome). Shelford first proposed the concept of biome. V.E. Shelford, a pioneering figure in ecology of the first half of the twentieth century, was an animal ecologist, who conducted research into insect adaptation to climatic stress. His academic life was based in the Mid-west of the USA. As a graduate student he was much influenced by the ideas of Cowles, and in later life he corresponded with Tansley. Thus he was interested in the developing concepts of succession and the ecosystem. Shelford was also interested in the application of biological science to resource management. He corresponded with Gifford Pinchot, one of the founders of modern biological conservation, as well as being involved in advising on protection of natural areas in the USA. The ideas of Shelford link to an early idea about climatic classification. The Köppen system, which is described in Box 8.1, was first

---

Box 8.1

The Köppen climatic classification

This system is based on the identification of a series of average climatic parameters, which are the thresholds for support of large-scale vegetation units. Although predating Shelford's ideas about biomes, there are similarities between the two systems.

The system uses a hierarchy of subdivision of climate, denoted by letters. There are three levels. The primary level is based on threshold temperatures.

The primary division is by temperature conditions, with the exception of hot deserts, and five types are defined:

A – Humid tropical. No thermal winter, with all months having a mean temperature above 18°C.
B – Dry climates, in which evapotranspiration substantially exceeds precipitation over the year, and there are constant water deficit conditions.
C – Humid middle latitude climates with mild winters. These are defined as having a coldest month with an average temperature below 18°C, but above –3°C.
D – Humid middle latitude climates with cold winters. These are defined as having a coldest month with an average temperature below –3°C, but the average of the warmest month is > 10°C.
E – climates with no thermal summer. These are defined as having an average temperature in the warmest month < 10°C.

These five primary classes are subdivided into fourteen subgroups, ten of which are subgroups and the remaining four are 'stand-alone' types.

For example, Af types of climate have at least 6 cm precipitation each month and Am types have a short dry season in which precipitation is < 6 cm, but is a defined minimum proportion of annual rainfall. In the E category there are two types: T with an average temperature in the warmest month of > 0°C but < 10°C; and F with an average temperature in the warmest month of < 0°C.

The third level involves further subdivision based on further specification of temperature conditions. For example, in C and D types the letter *a* indicates climates in which the temperature of the warmest month is > 22°C, and there are four months in which the average temperature is > 10°C.

For a fuller description of the Köppen climatic classification system, see Tarbuck and Lutgens (1997: 465–9).

devised in 1884, and, with some refinements is still one of the most widely used global climatic classification systems. It is based on temperature and moisture thresholds, which are regarded as the key determinants of broad types of vegetation. Shelford's biomes were essentially climatically determined, and relate to the Köppen system.

This view of biological communities represents the top of a hierarchy of biological organisation. This hierarchical perspective has been criticised for much the same reasons that the basic notion of phytosociological units, such as associations, have been criticised. But in the same way, as was indicated in Chapter 6, and illustrated by the NVC system, there is a fundamental value in classification, particularly for practical purposes. Provided its limitations are recognised, the biome approach is a good way of looking at the patterns of variation of communities in space throughout the biosphere. Furthermore, the biome approach can be linked directly to both the ecosystem and to functional ecology. Primary production and primary producers, that is, autotrophic plants, have the central role in determining biome characteristics and their spatial extent.

The environmental factors that control primary production and the global pattern of terrestrial biomes

The environmental factors, which influence large-scale patterns of primary production, have already been mentioned in Chapter 3 (Energy flow and energetics). Primary production, that is the outcome of photosynthesis, is the main function of vegetation. Rates of primary production vary from 0 to about 85 kJ/m^2/year, the former on the interior ice-cap of Antarctica and the latter in tropical rainforests, estuaries and coral reefs. The range of primary productivity found in biomes and major marine ecosystems is shown in Table 8.1.

Table 8.1 Global patterns of gross primary production

Biome	Area (millions of km^2)	Gross primary productivity (J/m^2/year)	Productivity index* (Hot desert = 1)	Approx. total gross primary production (J/year × 10^{15})
Open ocean	325.0	4,200	5	1,365
Coastal areas	35.0	8,400	10	285
Ocean up-wellings	0.5	25,000	30	10
Estuaries and coral reefs	2.0	85,000	100	160
Tropical rainforest	13.0	85,000	100	1,135
Temperate forest	5.0	32,500	40	125
Boreal forest	10.0	12,500	15	100
Grasslands	40.0	10,500	12	45
Low intensity agriculture	10.5	12,500	15	130
High intensity agriculture	4.5	50,000	60	200
Deserts and tundra	42	850	1	3.5

* Index of gross primary production is the factor for primary production in the biome/ecosystem, relative to that of lowest productivity biome, hot deserts = 840 joules/m^2/year. Thus open ocean productivity, which is approximately 4,200 joules/m^2/year, has an index of 5.

The environmental factors controlling rates of primary production are obviously those that directly influence photosynthetic rates. These are the availability of material inputs – nutrients; the availability of solar radiation – light; the heat environment of the eco-systems, as metabolic rates are generally influenced by temperature; and human impacts on biota and their environment. Material inputs are water, carbon dioxide and mineral macro- and micronutrients. The latter must be in available water-soluble form, so this part of the input to primary production is controlled by the aqueous solution at the plant roots, or tissue in the case of aquatic plants. Even in water, in which dissolved atmospheric CO_2 (or bicarbonate ions (HCO_3^-), depending on pH of the system) is the source of carbon, supplies of CO_2 are generally sufficient for that rate of photosynthesis that is possible, give all other controlling factors (with the exception of certain oligotrophic lakes, as discussed in Chapter 4). In other words it is not generally a limiting factor in the sense of Liebig's Law (see p. 60).

At the surface of the biosphere there is generally plenty of light radiation for photosynthesis. This remarkable process is energetically inefficient, using only about 1.5 per cent of available radiation. The large-scale exceptions to this pattern, which again means that light availability is not a limiting factor, are as follows. First, polar areas located beyond the Arctic and Antarctic circles have periods of the year during which there is little or no light. Second, light is readily absorbed by water, so that below a few metres' depth in water there is insufficient light for photosynthesis. Thus most of the volume of the world ocean (> 99 per cent), that part below the photic zone, cannot support photosynthesis. Temperature varies in complex ways over the Earth's surface in both time and space. Photosynthesis is very difficult at low temperatures, particularly if temperature falls below 0°C. Water expands by about 10 per cent when it changes state from liquid to solid form, and thus water within plants upon changing into ice will cause massive physical damage to plant tissues.

The water story is even more complex. Although some plants can survive in very low temperatures, mainly through dormancy, very cold liquid water is not a great hindrance to photosynthesis. As the maximum density of water is at about 4°C, cold water in ocean sinks, and surface water remains liquid. Only in polar regions are there any significant areas of frozen ocean, where ice shelves overlie the sea. In open oceans close to the surface where light is available photosynthesis is possible. The reasons for the low gross primary productivity are not due to the coldness of much of the world ocean.

However, temperature is a significant control in terrestrial biomes. Over some 60 per cent of the Earth's terrestrial surface, air temperature will fall below 0°C at some time during the year. Generally no photosynthesis is possible in these conditions, and plants must be adapted to deal with this stress simply to survive. Dormancy has already been mentioned, but there are other strategies, including surviving below ground as a tuber or in seed form. An important element in this thermal regime is seasonality. This is a result of the fact that the Earth's axis of rotation around the sun is tilted at an angle of 23.5°, with respect to the Earth's own axis of rotation. This mechanism is clearly explained and illustrated by McKinney and Schoch (2003: 122–3). This pattern of seasonality where there are alternating thermal periods which are favourable and unfavourable for photosynthesis is a great advantage to life on Earth, extending the area of land surface on which some photosynthesis is possible by at least 30 per cent. Resultant vegetation patterns and biomes are characterised by adaptation to the alternation in temperature, and particularly the thermal 'winter'. Dealing with winter inevitably means lower rates of primary production when compared with the inter-tropical zone in which year-round photosynthesis is possible.

To those of us living in the middle latitudes, seasonality means the cold winter, but in biological terms, over much of the Earth's surface seasonality means something different. Here seasons are defined not by heat conditions, but by the availability of water. We may speculate why life has bothered to evolve on the land. After all, the sea is all water and seems to provide everything else needed for photosynthesis, and what is more these inputs are immediately to hand. But this is not the case. Something is missing for high levels of photosynthesis in the oceans: an adequate supply of macro-nutrients. Nutrients fall out of the photic zone in the form of detrital rain, particles of dead organic matter which fall to the abyssal ocean floor, which may be several kilo-metres below. Generally there is no large-scale mechanism to return these nutrients to the biosphere, except by up-welling currents in a few parts of the oceans, and sediment uplift over geological time scales. Much of the world ocean is desperately nutrient-poor, and this is the limiting factor over most of the world's ocean surface. This explains the relatively low rates of primary productivity shown in Table 8.1. Only where there are mixing mechanisms, or where water is shallow and sediments can be disturbed by sur-face wave action, is primary productivity higher. The cold polar seas are surprisingly productive because the limited range of vertical change in water temperature makes nutrient movement easier. In tropical waters a clear temperature barrier, formed by warm, less dense water overlying cold, dense water, the boundary of which is termed a thermocline, is the cause of permanently low nutrient status in the photic zone. If other inputs are available, generally the supply of macro-nutrients is much greater on land than in aquatic environments.

However, the big problem for plant life on much of the Earth's surface is the avail-ability of water. Water must be available in the rooting zone, generally in large amounts, as much more water passes through a plant than just the amount required as an input to photosynthesis. Water availability is related to the movement of water in different states in the Earth's atmosphere (Chapter 4). This is a key dimension of climate. Climate may be defined as a function of conditions of the lower atmosphere, including dimensions of average temperature, precipitation (e.g. rainfall), wind, pressure and so on. At a global scale there are fairly clearly established climatic patterns in both time and space. Response to the climatic environment with the opportunities and challenges that it pre-sents to plant life is the key factor in terrestrial plant primary productivity and thus the resultant pattern of biomes. It is the primacy of climate, linking heat and water avail-ability, that is the major control of the pattern of biomes. This global pattern is shown schematically for terrestrial primary production in Table 8.2 which depicts variations in production rates North and South of the equator. This shows that primary production decreases, moving pole-wards. On this global pattern is superimposed the continental scale patterns of variations in water supply.

Table 8.2 Primary production rates by latitude North and South of equator
This table gives generalised index values for terrestrial primary production, as there are considerable variations in primary production at any latitude depending on the environmental conditions (heat, water, etc.) that apply to a particular location. The index value of 10 applies to equatorial latitudes where primary production is about 2.0 kg of dry matter per m^2 per year

Latitude (N & S)	Range of primary production indices
0–20	10–2
20–40	8–0
40–60	4–1
60–pole	4–0

Finally, it should be noted that biomes are affected by human actions, and indeed have been for millennia. The impact of human actions is discussed briefly in this chapter, and analysed more fully, at different spatial scales, in the final three chapters (9 to 11). The impact of humans, even on these largest of biological units, is significant and increasing. This is a cause for concern, but we must remember that humans have had large-scale impacts on the biosphere since our species evolved. This does not make contemporary problems any less serious, but it does mean that these problems are not new. However, there is an overwhelming body of scientific evidence which shows that problems are now greater than at any time in history. Furthermore, the concepts of ecological stability, and related issues of fragility and non-linear dynamics, which were discussed in Chapter 6, are factors which make current acute problems of human impact even more critical.

It is not simply variations in total primary production that are controlled by climate at the largest scale. The functional ecology of primary producers is influenced by climate. Although at any location plants with appropriate functional strategies will exploit small-scale niches, there is a general tendency for particular strategies to be favoured in certain climatic environments. Thus in humid inter-tropical areas where all requirements for photosynthesis are relatively abundant, competitors tend to play dominant roles in climax vegetation. Where water shortages are significant, plants that can develop quickly and exploit opportunities in periods when moisture is available are important. Prominent among these are grasses, which have many disturbance-tolerant (R) characteristics. In the most extreme environments – dry, cold, or where there are extreme nutrient regimes – stress-tolerant species are important elements in the vegetation cover, which is inevitably limited.

The global pattern of biomes

In this section the pattern of distribution of biomes is assessed. The analysis of biomes is grouped into broad climatic belts. Spatial patterns are shown in a series of maps. The main characteristics and environmental controls for each biome are discussed briefly. Furley and Newey (1983, part 5: 223–358) give a fuller account of biomes. For each biome the extent and general characteristics of the dominant vegetation type is explained, followed by a short discussion of the whole range of biological communities and the relationships between the biome and world-scale patterns of climate.

The humid and sub-humid tropics

The biomes represented in this zone are tropical rainforests, including evergreen and seasonal types, savannahs and some grasslands. The distribution of these types is shown in Figure 8.1.

These types of vegetation include rainforests, which have the highest primary productivity, the greatest biodiversity and most structural complexity of any terrestrial biome. The richness in biodiversity extends to all trophic levels and all types of organisms, including microbiota. It has been estimated that only a small proportion of species that actually exist in rainforest ecosystems have yet been identified. For example, Lévêque and Monunoou (2003: 23) state that while about 950,000 species of insect have been identified, there are probably some 8,000,000 species in existence. The majority of these unidentified species are in the humid tropics. Where water is abundant a profusion of different species of large, tall trees (see Chapter 7) dominate the vegetation. The forests are incredibly species-rich, and competition for light among the tall trees is intense.

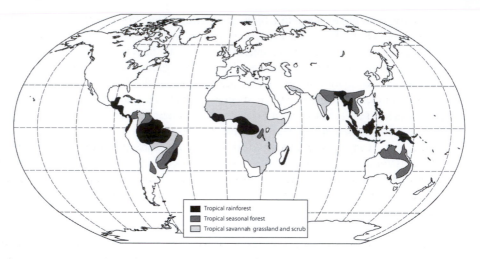

Figure 8.1 Tropical forest, savannah grassland and scrub biomes

Most primary production is concentrated in the many species of tall trees. However, there are many other plant species adapted to dealing with the overwhelming competition for light, water and nutrients that the big trees exert. The three-dimensional mosaic of plant species, large and small, is the driver of both the very high rates of primary productivity and the high indices of biodiversity.

As rainfall decreases, the characteristics of rainforest change, but the vegetation type remains that of the rainforest biome. What have come to be thought of as classic rainforests, equatorial evergreen rainforests of humid tropical areas, becomes modified into seasonal and other types of rainforest, and are located in areas further away from the equator, in which there are dry seasons. Whitmore (1998: 14) has developed a typology of forest types, using climate, soils and elevation as variables. Seasonal rainforests, such as the remnants of the vast forests of southern Brazil, have only somewhat lower biodiversity and productivity than equatorial forests. Much of these have been lost by conversion to cultivation and grazing land, though not all of this loss can be ascribed to the modern period. Protection and restoration of these forests is a high priority for conservation in the inter-tropical zone.

Primary consumption in the rainforests, effectively 'grazing', is mainly carried out by invertebrates. These attack not only the nutritious green tissues of the tree canopy but all areas of the sub-aerial parts of the forest vegetation. The height of the canopy makes it difficult for larger primary consumers to play a dominant role in grazing, and many of these are secondary consumers. Food chains are long and complex, with many organisms consuming at different trophic levels. The role of the prolific detrital chain in the soil is vital. Climatic conditions are ideal for the operation of a vast range of micro- to small biota, from bacteria to insects. A complex sub-surface food chain develops here too. Much of this life remains unrecorded. Some organisms that exist in enormous numbers, such as termites, are adapted to break down the resistant lignified tissues of the tall trees. The outcome is rapid and very complete nutrient cycling. This provides nutrient input for the productive forests. The rapidity of cycling and immediate uptake by plants means that the soils of these forests are nutrient-poor in terms of available nutrients. This paradox, the contrasts between the luxuriant forest cover and the infertile soils, is a key indicator of the fragility of this ecosystem. Breaking the rapid cycling systems, by forest clearance for agriculture, has been the cause of extensive and profound land degradation throughout the tropics.

Current ecological thinking is that throughout the world tropical forests have had a more dynamic history than was once thought to be the case. Stratigraphic, palynological and analysis of other biological micro-fossil evidence shows that climate change, particularly in the late Pleistocene, and early human impacts produced vegetation conditions during the past 100,000 years which were at times quite different from the 'classic' virgin rainforest, in areas which had previously been believed to be continuous climax rainforest for up to a million years. This puts an interesting perspective on conservation efforts for contemporary tropical rainforests. There remain many gaps in our knowledge of this most complex, biodiversity-rich and productive biome. For example, the concept of refugia or refuge zones to which tree species could retreat during periods of climatic deterioration, and from which they could spread in a subsequent amelioration, is being challenged by stratigraphic and micro-fossil evidence from tropical forests in both the old and new worlds.

The rather less productive (though still biologically diverse) savannahs and sub-humid grasslands have sufficient productivity to support the greatest concentrations of large grazing animals. Savannahs and grasslands of the sub-humid tropics have an enigmatic status. It is generally agreed that fire used by early humans has played a role in current vegetation patterns. Climate change is also a factor, and current moister climatic conditions might support more tree cover than is now found in savannahs (Goudie 2000: 62–7). Savannahs and grasslands contrast strongly with rainforests in two ways other than the obvious factors of lower primary productivity and dominance of most of the vegetation cover by grasses. Soil has more available nutrients. Although breakdown of litter is rapid and complete, uptake is less swift. Although rarely very nutrient-rich, savannah soils are less fragile than rainforests. However, overgrazing or inappropriate cultivation by humans can lead to degradation, and in severe cases to **desertification**, as is discussed in Chapter 10. Second, grazing in savannahs may be dominated by large vertebrates. The accessibility of grasses to such animals is the key element in primary consumption, though invertebrates still make a significant contribution to grazing. Secondary consumers, such as big cats, form the top trophic level. As large animals, evolved to deal with their large prey, they require extensive hunting territories. Conflict with humans is long-standing, and habitat destruction, which restricts territories, is the chief threat to these animals. Their conservation status is high partly due to their endangered status, but also because of the positive way in which humans regard these creatures. The term 'charismatic megavertebrate' applies very well to the lions and cheetahs of the savannahs of East Africa.

Hot deserts – desert and semi-desert

About 30 per cent of the Earth's surface falls within what has been termed 'the arid zone'. This includes semi-arid rangelands, semi-deserts, deserts and hyper-arid deserts. This range of types follows decrease in rainfall. The pattern of biomes associated with these types of arid land is shown in Figure 8.2. Hyper-arid deserts generally have extensive areas that are devoid of vegetation. Primary productivity is close to zero, and there is very little life present. It is notable however that even in such areas there may be some dormant life in the form of the plant seed-bank in the substrate. In some rainless areas sub-surface water movements may allow highly drought-resistant plants to survive in a few locations. Such locations are more extensive in deserts in which there are occasional rain events. Channel-like valleys such as canyons and wadis, which may be of ancient origin as well as influenced by current fluvial action, may have sub-surface water beneath the sediments that make up their flat floors. This water is present as a result of infrequent

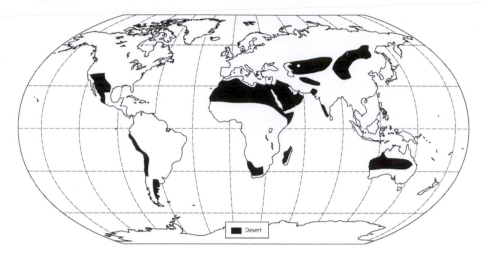

Figure 8.2 Desert biome

rain events, or in the form of underground resources. These latter may be the result of contemporary sub-surface water movements, or 'fossil', having origins in rain that fell in the past, when different climatic conditions occurred in the area. This habitat supports stress-tolerant plants, which can deal not only with scarcity of water, but with water of high salinity. The salinity of soil water is a function of the extremely high rates of evapotranspiration. In extreme cases soil pH may exceed conditions in which plant growth is possible.

Biomes of the arid zone are characterised by low levels of primary biological productivity. This is the inevitable result of scarcity of water, a continuous supply of which is needed for photosynthesis. Plants develop functional strategies to deal with water scarcity. These are summarised in Table 8.3, and are discussed in more detail in Chapter 5. In spite of low total primary productivity, the biodiversity of these biomes is not as low as might be expected. The range of micro-habitats in desert conditions gives opportunities for a range of stress-tolerant plants (and animals making use of the plants) to exploit these. Although there is competition for water, the differing habitats, for example, with different chemical water quality, and different amounts and depths of water available, mean that highly specialised plants can exploit particular habitats. The higher trophic levels are inevitably influenced by the harsh thermal environment and small amounts of energy available for consumption. There is a tendency for biodiversity to be at its highest in locations in which there are some limited water supplies, and for biodiversity to decrease as water availability increases, and more competitive plants can be sustained. An example of this pattern, from the eastern desert of southern Egypt, is considered by Ali *et al.* (2000).

At higher trophic levels in desert biomes, animal occurrence is controlled by the harsh climatic environment and low levels of energy available. Large animals are scarce. Consumers, both primary and secondary, often have large territories. Many animals are adapted to dealing with extremely high surface temperatures, which may exceed 70°C, by such adaptations as nocturnal activity, or living most of the time beneath the soil surface. A high proportion of life is poikilothermic, including reptiles and numerous insects and arthropods. Adaptations to allow animals to live with a very limited input of water, and to be able to cope with long periods without water, are found widely among desert fauna (see Chapter 5 for more on desert animal adaptations to stress).

Table 8.3 Plant strategies in drought conditions

Strategy	Adaptation	Size	Life cycle	Water requirements
Escapers (ephemerals)	Go through life cycle very rapidly (a few weeks), and survive dry periods as seeds which may be viable for many decades	Small	Very short (a few weeks)	Need continuous water supply during germination period
Endurers (true xerophytes)	Slow but continuous growth	Small to large (trees)	Intermediate (several years), to very long (centuries)	Need water more or less continuously (may be of poor chemical quality and at depth > 10 m)
Evaders (periodic growth xerophytes)	Slow growth during favourable periods, in which some water is available	Small to medium (small bushes)	Long (up to several decades)	Use water when available; typical of areas in which there are some infrequent rain events
Resisters (cacti and succulents)	Slow growth and specialised water storage and conservation mechanisms; generally very low levels of photosynthesis	Small to large (largest cacti are tree size)	Long to very long (many decades)	May survive extended periods (> 10 years) without water

Scarcity of water is the limiting factor. Humans have tried to offset this since the dawn of civilisation. Indeed, it is notable that among the earliest so-called 'hearths' of civilisation are the desert areas traversed by great rivers (e.g. the lower Nile and Mesopotamia, the zone between the Tigris and Euphrates rivers in present-day Iraq). Here some of the first plant domestication took place. The light, easily cultivated soils provided an ideal habitat, given water, for disturbance-tolerant grasses, which are the ancestors of our most important cereals. Water was available from the great rivers. Natural flood regimes were soon supplemented by basic irrigation mechanisms. In the twentieth century, huge areas of the margins of the Sahara, the deserts of Central and South Asia and in the southwest of the United States, as well as many other parts of the world, have been made into highly productive, intensive agricultural areas by large-scale irrigation schemes.

However, irrigation comes at an environmental cost. As more water is available in arid areas, so evapotranspiration increases. Soluble salt in the soil and substrate is drawn to the surface and precipitated at the soil surface. This process in extreme cases will result in salinisation of the soil, increasing pH to > 10, and rendering the area effectively sterile. Once such conditions are in place they are very difficult to reverse, and the result is permanent and severe land degradation. Very careful management of irrigation water, and good sub-surface drainage, a condition that is frequently not the case naturally in arid soils, are essential to prevent salinisation. In most large-scale irrigation schemes of the twentieth century at least 20 per cent of the irrigated area has been affected by salinisation. The problem is found in both more and less economically developed areas. Nor is the problem new. More than 6,000 years ago the Sumerians of Mesopotamia developed ingenious irrigation systems based on qanats, interconnected wells. Archaeological evidence has revealed that these irrigated lands were affected by salinisation.

The middle latitudes – deciduous forests, temperate grasslands and Mediterranean biomes

Much of the developed world is located in this climatic region. It is perhaps that group of biomes most modified by humans, and conversely least of which in anything approaching a natural state remains. Figure 8.3 shows the extent of the natural range of the biomes of this area. However, over most of the world all parts of the ecosystems of these middle-latitude areas have been modified or replaced by agricultural or sylvicultural land use. A small but significant proportion has also been lost to urbanisation and industrial activities. Deciduous forests once covered most of the more humid parts of this area. Very little natural woodland remains in Europe and North America, though there are somewhat larger areas of semi-natural woodland, which have been planted or managed by humans. The characteristics of these forests contrast with tropical rainforests. Temperate deciduous forests have significantly lower primary productivity, about 40 per cent of that in tropical rainforests. This lower level of productivity is largely caused by dormancy of trees during the cold winter. Biodiversity is lower and vegetation structure less complex. Soil conditions too are different. Lower rates of decomposition and uptake mean that the proportion of plant nutrients which are in available form in the soil is smaller. Thus in human terms deciduous forest soils are fertile. This accounts for the reason why so much of this biome has been converted into intensive forms of agriculture. Species in the higher trophic levels are far less numerous than in the tropical forests.

Temperate grasslands, prairies and steppes are semi-natural grasslands located in drier parts of the middle latitudes. As with tropical savannahs, human actions, particularly the use of fire and grazing, have played an important role in shaping and maintaining the

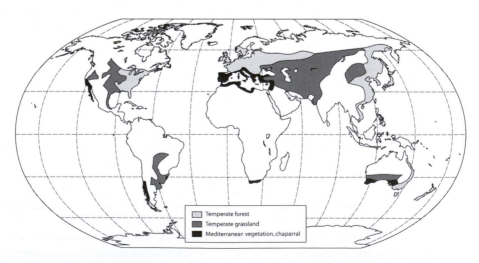

Figure 8.3 Temperate forest, temperate grassland and Mediterranean biomes

ecological characteristics of these biomes for millennia. Like temperate deciduous forests, much of these areas have been converted into farmland both for growing cereals and for extensive range farming. In more advanced countries, irrigation is being used to protect and boost yields of crops in the drier parts of these grasslands, where they merge into semi-deserts. Burning may still be used for rangeland management. Introduced species, such as weeping lovegrass (*Eragrostis curvula*), a domesticated grass originally from southern Africa, which provides good grazing for cattle, are now grown extensively in the pampas grasslands of South America. The use of fire and problems of introduction are discussed in Chapter 9.

Mediterranean climates are an extreme variant of the temperate zone in which a long, hot summer, caused by effectively tropical climatic conditions migrating seasonally, alternate with cool and moist winters. The alternation between water surplus and deficit, the latter through much of the growing season, has produced a very distinctive type of biome. Most vegetation has functional strategies that deal with the summer drought. Xerophytic adaptations are common. Open forests and scrub have been much modified by humans through fire grazing and cultivation. Semi-natural vegetation formations with distinctive plant assemblages, but possessing a common functional ecology, have developed in Mediterranean climate zones in different parts of the world, known as maquis or mattoral in Europe, chapparal in North America and mallee in Australia. Pressures on the remaining areas of Mediterranean forests and scrub are great. Human pressures on these areas have been substantial since classic times and, particularly in mountain areas, soil erosion and land degradation are continuing problems.

Throughout much of the part of the world occupied by temperate biomes, and that are home to the majority of people living in more economically developed nations, the notion of 'land improvement' has been a major cultural theme in the economic history of these nations. Historians have analysed the extraordinary efforts of Europeans and their colonial descendants to bring 'civilisation' to all parts of the world (see e.g. Ferguson, 2003). In this endeavour the colonists were exporting their cultural and ethical values, and trying to re-create the same taming of the wilderness that was responsible for agricultural improvement at home, itself the necessary precursor of the Industrial Revolution and the power which went with it. Indeed, these actions were seen by some through notions of the 'Protestant work ethic' and similar concepts to be divinely ordained and the sacred duty of humans. The actions and ecological changes that followed have had profound effects on human history and on world ecology, and were largely driven by experience in the temperate biomes. We cannot explore the historical perspectives or ethical viewpoints relating to European influences here, but we must note that it is a very important worldwide ecological factor, and understand its ecological roots.

The basis of the changed ecology of the middle latitudes was and is intensive agriculture. It has a particular functional ecology based on control and simplification of ecosystems. In the eighteenth century in Britain the agricultural revolution, or more properly the second agricultural revolution, distinguishing it from the first Neolithic event, saw food production soar as labour inputs into agriculture fell. This was achieved by enclosure, rotation, use of fertilisers, mechanisation and new strains of crops and animals. The ecological strategy was to simplify and control ecosystems so that the crop or output from the system to be used directly or indirectly by humans was maximised. Yields were boosted by eliminating competitors (weeds) and pests (secondary consumers). The fertile and resilient soils of the temperate zone could sustain the new agriculture indefinitely if the new methods were applied wisely. Modern agriculture has almost entirely abandoned sustainability, using instead material and energy subsidies to substitute for natural processes. In terms of primary productivity this system is successful, with rates higher than that of the natural system it replaced and about 60 per cent that of tropical rainforests. This is only possible, however, through large-scale use of material and energy subsidies. Furthermore, attempts to export such agricultural methods to other environments, particularly the fragile biomes of the tropics, have often been very damaging environmentally and unsustainable. These actions have resulted in many environmental and ecological problems, some of which are discussed in Chapters 9 to 11.

However, in parts of the temperate zone, humans exist in closer harmony with nature. Significant areas of Europe and North America have protected status, in which the priorities for management of ecosystems are conservation and recreation. Impacts upon, and damage to, ecosystems if not eliminated are reduced. The extent of such areas and their importance to the populace is growing. Furthermore, for many centuries some areas have been used less intensively for grazing land. Pasture and heath managed by grazing and burning are important refuges for all kinds of wild biota. Again the perceived significance of such low-intensity systems is growing, and the future will most probably see elements of ecosystem management in which conservation is in partnership with agriculture. A similar scenario may be seen in the replacement of intensive sylviculture ('factory farming of trees') with a more natural and ecologically attractive forestry system. In such forests, native deciduous trees are the main species used, rather than exotic, fast-growing conifers, and the resultant woods have a more open and diverse structure, with trees at different stages in their life cycle present everywhere.

Continental interiors of middle latitudes and coasts of the subarctic

In parts of the middle latitudes, extending in the northern hemisphere to 60° north, winter climate is severe, with temperatures below 0°C for extended and continuous periods, and with short though warm summers, in some cases with little more than 120 days when temperatures are above the threshold for growth. Typically these areas are found in the interiors of continents, far from oceanic moderating influences. Most such areas are quite dry and some merge into semi-desert, though generally there is enough rainfall to sustain tree growth. However, primary productivity is low, typically less than one-sixth of that in tropical forests. Correspondingly, biological water demand is much lower and evapotranspiration rates over much of the year are low. The limiting factor controlling vegetation here is the short thermal growing season. Similar thermal though moister conditions may be found in the coastal areas of the subarctic, such as southern Alaska, southern Chile or western Scandinavia. In all of these locations conditions are ideal for trees, particularly of a handful of species of large, slow-growing conifers. This is the boreal forest biome, the extent of which is shown in Figure 8.4.

Boreal forests are dominated by a handful of species of conifers, typically pine (*Pinus*), spruce (*Picea*) and larch (*Larix*) species. Often huge stands of thousands of hectares' extent are composed of one or two species of tall tree. Structure is simple, with often only two or at most three layers of vegetation, and biodiversity is low. Both plant and animal life must be adapted to the long, cold winter. Temperatures may fall below −25°C. At such levels plant growth is impossible, and animal life retreats into dormancy or migrates to more equable climes. In contrast, the short summers, during which average maximum temperatures may exceed 25°C, see a burst of frantic biological activity. Everything is done in a rush, because it has to be, to beat the onset of the long winter. The few tall tree species, which utterly dominate the biome, grow slowly, not reaching maturity for twenty years and living for a hundred years or more. Primary consumption is dominated by birds and insects in the tree canopy, and a few species of large grazing animals use the forest floor for summer grazing. In the winter, the birds and mammals migrate or become dormant, together with their predators, while the prolific summer

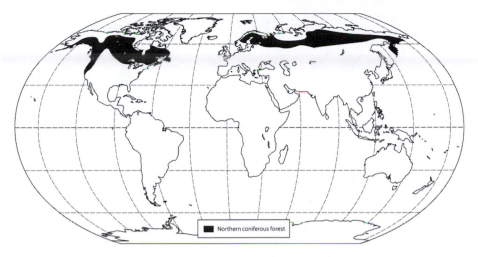

Figure 8.4 Northern coniferous forest biome

insect life overwinters as eggs and pupae. In all cases the primary functional ecological strategies are based on coping with the thermal stress of winter.

The soil and decomposing systems in boreal forests are quite different from those in both tropical rainforest and temperate deciduous forests. There are large amounts of unavailable nutrients in soil litter. This is related to very slow rates of action in the decomposing system. Not only is decomposition impossible during winter, but the high carbon/nitrogen ratio of the plant litter means that organic debris breaks down into acidic humus. Such material, with a pH of 3.5 or so, prevents bacterial action, and so decomposition is carried on by fungi and actinomycetes. These act slowly and breakdown is often incomplete. Thick, peaty layers are common in the surface of soils under boreal forests. In many ways the boreal forest soils are the opposite of rainforest soils. The latter have rapid and complete recycling and uptake of nutrients, while in the former recycling is incomplete and very slow. The similarity between the two is that in both cases soil is poor in available nutrients. Thus both contrast, for different reasons, with the nutrient-rich soils of the temperate middle latitudes. Boreal forests are not immune to human impacts however. The trees themselves are valuable raw materials which have been ruthlessly exploited. The long life cycle of this biome means that natural regeneration may take several decades. Even though cycling systems are slow, breaking them may result in environmental damage, particularly in wetter areas and slopes where soil erosion is an almost inevitable consequence of clear felling.

Arctic and mountain biomes

The axial tilt of the Earth, the cause of seasons, allows life to penetrate to regions close to the poles. No plant life is possible on ice-caps but around their margins is the near-treeless tundra. Similar vegetation types are found in mountain areas above the tree line and below the permanent snow line. These are various categories of mountain heath. The extent of these biomes is shown in Figure 8.5. Note how most of this biome is located in the northern hemisphere, as there is little land mass within the corresponding climatic zone in the southern hemisphere. Primary biological productivity in these biomes is at about the same level as terrestrial deserts or oceanic waters. Only a few species of small plants can survive the cold climate in which temperatures in the short growing season

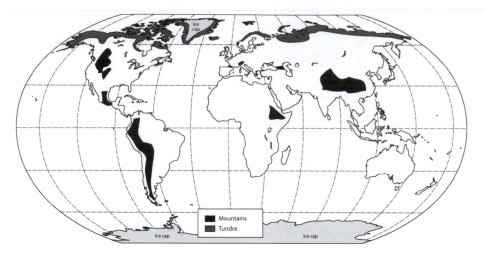

Figure 8.5 Mountain and tundra biomes

rise little above the threshold for growth. Average summer maxima may be only 12 to 15 °C. Vegetation is dominated by hardy grasses, mosses and a few shrubs. Permafrost (discussed in Chapter 6), and cryoturbation, substrate mobility caused by freeze–thaw action, makes the rooting zone a highly stressed and disturbed environment, in which only species able to cope with this environment can live. There may be a fair amount of insect life in the short summer and a little grazing by migratory animals, but overall very limited biodiversity matches and is explained by the low levels of primary productivity.

However, as with boreal forests, this is a fragile environment and its ecosystems are easily damaged by even low levels of human impact. Although there is little to interest humans in direct harvesting of biological resources, exploitation of underlying minerals is a serious issue. In areas such as the north slope of Alaska or northern Siberia, where there are large reserves of oil, a critical, increasingly valuable and scarce resource, the potential impact on ecosystems is severe. Development of the resources here has to be carefully planned, to avoid damage to substrate and allow for animal migration. The attraction of these remote areas to the inhabitants of the crowded rich cities of the more developed world lies in their remoteness and pristine natural condition. In a few parts of these biomes the pressure from visitors is increasing to the extent that access to the most vulnerable tundra and mountain heaths has to be controlled. This situation is likely to become more common in future.

Wetlands and freshwater aquatic ecosystems

Throughout the world, the ecosystems which occupy freshwater bodies, lakes and rivers, and their margins, are distinctive. Although there are differences in species composition and biodiversity, related to climatic conditions and water chemistry, there are common functional elements in the ecology of freshwater bodies and their margins, which permit a general analysis of these ecosystems at a global scale. Thus freshwater and ecosystems may be considered as a biome with a widespread but discontinuous spatial distribution. There are two broad types of primary producers in freshwater bodies. These are algae and other simple microscopic plants, free floating in water, **phytoplankton**, or attached to the substrate or larger plants (**phytobenthos**), and macrophytic vegetation, mostly vascular plants, some of which may be several metres in length and which may be fixed to stream or lake floors or free floating. These latter are aquatic macrophytes. Rates of primary production in freshwater are variable, ranging from virtually zero to levels similar to those in the most productive systems. The reasons for this variation lie in the controlling physical environment of aquatic biomes.

The dominant factor, which controls rates of primary biological production, is the availability of nutrients in water. Generally freshwater is nutrient-poor. In higher latitudes it is often very nutrient-poor or oligotrophic. In the inter-tropical zone, where rates of chemical action are increased as a result of higher ambient temperatures, nutrients may be more plentiful. The substrate over which water flows to the water body or river is an important factor influencing the nutrient content of water. Nutrients sink to the base of a water body, and are not available to primary producers in the photic zone. In temperate areas a seasonal rhythm of phytoplanktonic activity is related to transfer of sediment-rich bottom waters to the photic zone. These processes are discussed by Colinvaux (1993: 544–65). As is discussed more fully in Chapter 10, artificial nutrient enrichment of water by humans is an increasingly important issue, which affects both the quality of water resources and the conservation of freshwater bodies. Water temperature has an effect on primary production but is generally much less important than nutrient status. Light availability affects primary production through latitude and season to a very limited extent. Much more important is the turbidity, the transparency of water, which is related to sediment content and has a much more potent effect. In very turbid water no photosynthesis is possible more than a few tens of centimetres below the surface.

The higher trophic levels in freshwaters, including benthic invertebrates and bacteria and other macrobiota in bottom sediments, as well as fish, zooplankton and insect larvae, play an important role in nutrient cycling in these systems. In tropical countries where the nutrient status of water can support sufficient primary production to sustain large fish stocks, these latter may be important elements in human food resources. In temperate rivers and lakes fish stocks are valuable recreational resources.

Transcending this in human significance, the importance of life in freshwater bodies is in its role in abiotic cycling systems. Not only are there links with terrestrial systems, but also the cycling is vital to water quality. Consumer organisms can potentially make use of both autochthonous and allochthonous energy sources. The former is material that has its source in primary production of the water body, while the latter includes litter and other organic debris which enters the water from adjacent terrestrial systems. In nutrient-poor water the latter is vital, and makes a significant contribution where fringing vegetation is productive. This is one direction of the link to terrestrial biomes. The reverse flow is carried out by consumers, amphibious animals and fishing birds, which live on land but feed in water.

Life in water is vital for maintaining its chemical condition. Without consumption and decomposition waters would become loaded with nutrients and unusable to humans. Life in water is the safeguard of its purity. This has been recognised in Europe with the adoption of ecological standards for assessing water quality and guiding its integrated management. The Water Framework Directive (WFD) of the European Union provides the basis for future freshwater standards throughout Europe. It has been described appropriately as the most important piece of environmental legislation in Europe of the past fifty years, as its requirements will affect the ways in which most human activities are carried out. The WFD is described in more detail in Box 8.2.

Wetlands – fens, bogs, swamps and flushes – are terrestrial environments that merge with and link to freshwater ecosystems. The nutrient status of these areas is important. In some cases there are higher amounts of nutrients, though this is much less common than impoverished nutrient conditions. An example of the former is hill flushes, which form visibly green paths running downhill on the mountains of the highlands of Scotland.

Box 8.2

The Water Framework Directive (WFD)

The actual document, which explains this in detail, is 152 pages in length and imposingly entitled 'Directive of the European Parliament and of the Council 2000/60/ec establishing a framework for community action in the field of water policy'. At the beginning of this legal charter however this key phrase explains what it is all about: 'Water is not a commercial product like any other but, rather, a heritage which must be protected, defended and treated as such' (p. 3).

Under the WFD the assessment of water quality is based on ecological status of water. Reference standards of good (and lower) status water are defined. The objective of the WFD is to make all water bodies, including coastal waters, of good ecological status. This is a complete change from earlier physico-chemical stand-ards (e.g. water temperature, chemical content) and is a much more powerful tool for management. In the WFD management is to be integrated, meaning that *all* human activities, which influence or might influence water quality, are to be examined in ensuring that a good status for all water in EU countries is achieved or maintained.

Furthermore, the WFD uses **the precautionary principle** (and polluter-pays principle) as elements in developing water management strategies. All EU countries are governed by the requirements of the WFD, and in most European countries this has meant undertaking considerable engineering and scientific work to ensure that the standards are met.

However, the waterlogged substrate is generally inimitable to decomposition. Anaerobic conditions mean that the breakdown of litter is incomplete. Peat, semi-decomposed organic debris, may accumulate rapidly. Formerly seen in negative terms by humans, these areas now have high conservation significance. The so-called 'Flow Country' of the far north of Scotland is one of the world's greatest extents of peatland. Not only is it now recognised as having international conservation significance for birds and insects, but it also has global importance in carbon sequestration; that is, it holds organic carbon in a stable non-gaseous form. Its stability is important. If peat is abstracted, for use as a fuel following drying, or is desiccated by changes in the water regime, the carbon con-tent will be volatilised as CO_2, thereby accelerating the greenhouse effect. Plants living in wetlands are generally adapted to dealing with poor nutrient status.

Marine ecosystems

Life in the world's oceans is so complex and varied that it would fill a whole library. The following summary picks out some of the key elements only. Those readers who wish to know more can make a start by consulting the further reading recommendations at the end of this chapter. The open oceans have very low primary productivity. There is a con-trast between the middle latitude and inter-tropical deep oceans, which have very low, near desert productivity, with circumpolar oceans, and shallow seas which have higher primary productivity two or three times that of the open oceans. Photosynthesis is carried out by phytoplankton, of which there are relatively few major species given the vast areas involved. Photosynthesis is confined to the photic zone, which even under ideal circumstances does not extend much below 10 m in depth. In spite of the low rates of primary production, the vastness of the oceans means that the total amount of primary production carried out in the ocean is similar to the total of all terrestrial primary production. Primary consumption is the role of zooplankton. These tiny grazers are often the juvenile stages of larger organisms, and they in turn form prey to consumers of all sizes. Food chains are long, and the highest levels, such as sharks and some cetaceans, consume at the fifth level. That such huge creatures can exist in large numbers, and in the case of sharks have done so successfully for many tens of millions of years, shows how much total energy is available in the ocean, even if primary productivity rates are generally low. Debris of all types falls from the photic zone to the ocean floor. This is typically several thousand metres below the surface, and thus in permanent and total darkness. Here, strange-looking detrital bottom feeders live on this continuous rain of food. Humans still know relatively little about the ocean depths. For example, only in the past decade or so have the existence of ecosystems, which are based on the use of energy from hydrothermal vents, out-wellings of sulphur-rich hot water from volcanic vents, been discovered (see Chapter 3).

Some parts of the world's oceans have very high productivity These include up-wellings; that is, areas of vertical movements of water from the ocean depths, shallow seas and coastal areas in general, particularly coral reefs and mangrove swamps. The boost in productivity is due to the enrichment of water in the photic zone with nutrient-rich sediments from the seabed. Such areas team with life of all kinds. Coral reefs have biodiversity levels which are similar to rainforests, but mangroves, in spite of their productivity, are much less so, especially in plant life. The swamps are highly stressed environments as periodic inundation makes plant life impossible for plants which are not adapted to this regime. For those that are, for example, the mangroves, this permits very rapid growth. Consumers thrive on the energy available to them. These two systems show that there are no simple relationships between productivity and biodiversity.

Conservation of the oceans is just as important as that on land. The ecosystems of the oceans play a vital and equal role in all natural cycling systems. Our ignorance of and remoteness from the ocean tends to make us very careless about what we have done to it and to its life. We deplete its life by over-fishing and dump all manner of harmful materials in the ocean. We have assumed that the vastness of the ocean equates with a limitless ability to cope with all these threats. There is now a growing realisation that we cannot continue as we have in the past, and we must treat the ocean and its ecosystems in a sustainable way. What happens on land affects the ocean too, and not just what we dump in it directly. The effects of global warming may extend to the ocean through modification of ocean current systems, with dramatic effects on climate in Western and Northern Europe.

Variations within biomes, and the human factor in biomes

So far we have examined global patterns of ecosystems. We need to briefly examine smaller scale patterns and their causes. Many factors other than climate contribute to variations in vegetation and whole ecosystems, as was discussed in Chapters 2 and 3. These factors result in complex and continuous variations in life over the surface of the planet. In detail much of the cause is related to availability of water on land and nutrients both on land and in the sea. There is too much or too little for ideal plant growth, generally the latter, in many environments. Conditions can change rapidly in space, and also change over time at all scales from seasons to geological eras. In more general terms, at a meso-scale, there are a number of environmental factors which have widespread effects on biomes. Three examples are discussed here: it should be noted that there are many other such meso-scale factors.

Coastal areas and internal large water bodies experience a modified climatic regime as a result of thermal buffering by water, which has a very high **specific heat**. Added to this, the stressed environment of coastal areas means that climax vegetation in the successional sense may be different from that away from the coast. This effect is very strong within a few kilometres of large water bodies, but in a reduced form may affect a zone 100 km wide or more, where oceanic influences can penetrate inland easily, unhampered by mountains. Lakes have a much more localised climatic influence, though in the case of the Great Lakes of North America, the whole of the Lakes Peninsula of southern Ontario experiences a moderated climate. As a result, good-quality viticulture is possible here. Relief modifies climate considerably. Hills and mountains of the middle latitudes often receive wetter winds on their west-facing slopes. Leeward areas are not only drier but warmer due to **adiabatic** heating of air. This so-called föhn effect can produce significantly increased temperatures, for example, more than 10°C to the east of the Rocky Mountains, and up to 3°C to the east of the Grampian Mountains of Scotland. The latter are only about 1,000 m in height.

At the most detailed scales, down to a few metres in dimension, micro-habitat and niche are factors which modify the broad patterns of vegetation and the ecosystems that they support. This is largely a function of micro-climate, the localised effects of shade and shelter, often due to the influence of large plants such as trees. Changes in substrate geology and drainage may also be responsible for rapid spatial variations in vegetation patterns and ecosystem type.

Human impacts are increasingly important, severe and pervasive factors that modify ecosystems and their function. The effects are quite clear, even at the global scale of biomes. We see the outcome of human actions at biome level in deforestation and desertification. We also have evidence of the cause of some of these changes, for example, in rising levels of atmospheric CO_2. This issue is discussed more fully in Chapter 11. However, it is not easy to link these changes to what will happen in the future: both to climate and to the natural systems which are controlled by climate. Better data and modelling have improved our knowledge of what may happen in the future but knowledge is still imperfect. Some human actions, such as movements of people and overgrazing, may have a significant effect on ecosystem change that is difficult to separate from changes caused by climate change. Human actions result in changes to ecosystems, which are both deliberate (e.g. agriculture) and accidental (e.g. pollution). Changes may be deliberate or accidental as a result of some unforeseen accident or poorly understood chain of events. Changes to ecosystems and their functioning act at all scales right up to the global level. Finally, an important question which has more than philosophical significance is: Are human impacts on biomes and ecosystems different from other controlling factors? All of these issues are examined in the following three chapters. The issues we explore therein are the most important facing humankind in the immediate future.

Summary

This chapter covers:

- The nature of biomes, the climatically determined pattern of ecosystems over the Earth's surface.

- Biomes are effectively controlled by variations in climate, which in turn affect rates of primary production. The action of the factors that control primary production are analysed.
- The global pattern of biomes is described and explained.
- The patterns of variation within biomes at different spatial scales are discussed, and the effect of human actions on biomes is introduced.

Discussion questions

1   Which is the more important factor controlling primary production over the surface of the planet Earth: availability of light or temperature conditions?
2   Compare the proportions of nutrients in available and unavailable forms in the soils of tropical rainforest, deciduous forests of the middle latitudes and boreal forests.
3   The rate of primary production per unit area, that is, primary productivity, is very low in most open areas. Does this mean that such areas are unimportant ecologically and in terms of global primary production?

Further Reading

See also

Energy flow and energetics, Chapter 3
Material cycles in ecosystems, Chapter 4
Stressed ecosystems, Chapter 5
Disturbed ecosystems, Chapter 6
Productive and intermediate ecosystems, Chapter 7
Human impacts on ecosystems – impacts on trophic structure, Chapter 9
Large-scale impacts on ecosystems – the increasing effects of humans, Chapter 10
Global environmental change and consequences for ecosystems, Chapter 11

Further reading in Routledge Introductions to Environment Series

Environmental Biology
Natural Environmental Change

General further reading

Dynamics of Marine Ecosystems: Biological–Physical Interactions in the Ocean (2nd edn). K.H. Mann. 1996. Oxford, Blackwell.

Ecology of Aquatic Ecosystems. M. Dobson and C. Frid. 1998. Harlow, Longman.

Both of these books give more advanced treatments of aquatic ecosystems and their ecology.

Conservation Biology. A.S. Pullin. 2002. Cambridge, Cambridge University Press, ch. 2, 'Major world ecosystems', pp. 19–49. A short, up-to-date and nicely explained review of biomes and related issues.

Environmental Science: Systems and Solutions (3rd edn). M.L. McKinney and R.M. Schoch. 2003. London, Jones and Bartlett, pp. 106–10.

This gives a useful short review of aquatic ecosystems.

The Biosphere. I.K. Bradbury. 1991. Chichester, Wiley. Part 4, Spatial aspects of the biosphere, pp. 187–237.

This gives an interesting comparative approach to ecological zonation.

The Geography of the Biosphere. P.A. Furley and W.W. Newey. 1983. London, Butterworths. Part 5, The major biomes, pp. 223–358.

This remains one of the best accounts of world biomes.

9 Human impacts on ecosystems: humans as an ecological factor

Human impacts on ecosystems are as old as the human species. However, following industrialisation, with the consequent increase in numbers of people and their ability to modify the biosphere, both the extent and consequences of human impacts on ecosystems have accelerated. Impacts resulting from human activities occur in all parts of the biosphere, and at all kinds of temporal and spatial scales. This chapter covers:

- **General nature of human impacts on ecosystems**
- **Fire**
- **Introduced species**
- **Recreation**
- **Sustainable development**

Human impacts: an old and new issue

Human beings are part of the biosphere. In most parts of the world, humans are the dominant organisms. The previous chapters have shown that we share the biosphere with millions of other species. We also depend, as much as any other living creature, on the functioning of ecosystems in the biosphere to support our existence. Unlike all other species, people have the unique ability to affect profoundly the nature and functioning of ecosystems throughout the biosphere. This chapter is concerned with **anthropogenic** effects on the trophic structure and functioning of ecosystems. This is linked to functional ecology by examining the changes that take place in species composition in the affected ecosystems. In some ways humans may be considered as simply another biological species, albeit one that exists in very large numbers. But we are also the species that is capable of the most profound ecological and environmental impacts. The scale and importance of human impacts, together with the fact that (not unreasonably) humans tend to view the world from a human perspective, means that it is important to separate human roles in ecosystems from those of other species. There are now about 6,000 million individual *Homo sapiens*, and our numbers continue to grow. This huge population of rather large animals affects ecosystems, through elimination of species, modification of flows of energy or nutrients or by change to the abiotic environmental component of ecosystems. It not only affects all other species with which we share the biosphere, but also threatens the support systems for all of life on Earth, including that of humankind.

Human impacts on ecosystems have being going on since we first evolved. We should not think that it is only during the past 200 years, the period of human industrial societies, that significant impact on ecosystems has occurred. But it is true that the rate, scale and extent of change in the past two centuries have been much greater than what had gone before. This acceleration is a function of the geometric increase in the numbers of humans on the planet, and of the extraordinary increase in the ways and scale of change that this larger population has been able to undertake. Ecosystem impact on a major scale began when humans first used fire. Impacts accelerated with the domestication of plants and animals in the Neolithic agricultural revolution, and gathered further pace during the industrial and agricultural revolutions which began in Europe in the eighteenth century, and which have spread throughout the world during the following two centuries. Impacts may be deliberate or accidental. Most intensive agricultural activity is a deliberate attempt to modify ecosystem function for the maximum benefit of humans while nearly all pollution is accidental. Few humans actually want to foul their own nests.

It is difficult to develop general theories about impacts on ecosystems. However, impacts generally simplify ecosystem structure by elimination of some species or by modifying flows of energy and materials. Many impacts occur much more quickly than the ability of natural ecosystem functioning to restore the system to a similar state to that prior to that impact. As was discussed in Chapter 6, the fragility and resilience of the ecosystem affected significantly influence the outcome. Lags in system reaction follow-ing human actions mean that, within human time scales at least, change may be hard to reverse. In many cases it is impossible to return to the original state. Frequently, we do not understand the effects of impacts on ecosystems properly, and often we have little idea of the outcome of these changes. This can make **sustainable development** of natural resources, upon which the continued functioning of ecosystems depends, very difficult. Both poor knowledge and a low priority for ecosystem integrity remain as barriers to sustainable development. In this chapter we develop these themes using examples chosen to illustrate impacts acting at different spatial scales, at different rates, from different human origins, and which are located in differing types of ecosystems.

Fire

Since the dawn of human existence fire has been used deliberately to modify the ecology of different parts of the world. It has been a love–hate relationship. Fire can be and still is immensely destructive. Every year millions of hectares of productive ecosystems are lost to uncontrolled fires, and many human lives are lost too, in the countryside as well as in cities. Yet used wisely, and more importantly controlled carefully, fire can be a potent biological resource management tool. Initially fire was used by hominids, well before the appearance of *Homo sapiens*, as a hunting strategy. Large animals could be driven using fire into pits or over cliffs, where if they did not immediately die, they could be dispatched conveniently and safely. It is likely that the periodic use of fire for these purposes, allied to seasonal hunting territories, began the long process of domestication. Early humans might note that animals returned to newly burnt areas to graze the nutritious emerging grasses and other rapidly growing disturbance-tolerant plants. Soon animals too would become aware of this regular opportunity, and even though it brought them close to their ultimate predator, human beings, the reliable food source meant that that risk was worth taking. Furthermore, other predators were deterred by the presence of humans. This was the beginning of the symbiotic relationship between humans and domestic animals.

The next important deliberate use of fire was in land clearance. The easiest and quickest way of removing natural forest was to burn it. The open land could then be used for grazing. Later, as agricultural technology developed, this cleared land could be used for cultivation. In Europe clearance of natural woodlands proceeded rapidly from the beginning of the Neolithic period. By Roman times a significant part of the forest cover of the continent had been lost, and by the late medieval period the amount of forest cover was not much more than it is today. Some forest cover remained, as hunting reserves and in common lands and the like. This historical deforestation process has not only shaped European landscapes and their ecosystems, but also, as previously noted, conditioned European perspectives on use of forested lands. It is chastening to realise that we Europeans have carried out, most thoroughly and over many centuries, a policy of forest clearance to permit intensive agriculture. These are exactly the land-use policies we are now advocating that nations in the inter-tropical zone should *not* adopt. The agro-ecosystems, which replaced the forest, have a primary productivity roughly the same as in natural woodland, but this is at the expense of biodiversity and includes the cost of large energy and nutrient subsidies. Important small-scale habitats in agricultural land, helping to offset the loss of woodland, are hedgerows and shelterbelts. The ecological characteristics and significance of these are discussed in Box 9.1.

In tropical forests fire has also been a long-established management tool. The practice of slash and burn has been used for hundreds of years. Plots are carved out of mature forest by cutting down trees and burning out stumps and understorey vegetation. The nutrient-poor tropical soils are temporarily enriched by ash, and crops can be grown for typically up to five years. As yields fall and the forest encroaches on the plot, a new area is cleared and the old plot is left to recover and regenerate. The subsequent forest, dominated by pioneer tree species, is usually different from the original forest. Full recovery depends on the quality of the remaining seed-bank and the close-by presence of mature climax tree species. In most cases full recovery is unlikely without human intervention (Whitmore 1998: 151–2). However, with the above qualification, this system is more sustainable than permanent clearance for grazing cattle or growing soya, the fate of much Amazonian rainforest, millions of hectares of which have been burnt out over the past four decades. These immense fires are visible from space, disrupt air travel and increase CO_2 input into the atmosphere, as well as destroying irreplaceable rainforest and its treasure house of biodiversity.

When used carefully fire can be a useful grazing land management tool. The general principles are to remove old woody or fibrous vegetation by a controlled burn, thus allowing regrowth of young palatable and nutritious plant material. A controlled burn is one where the fire temperature is held at a level which will consume above-ground material, but will not destroy topsoil or the underground parts of plants. This will allow rapid regrowth from rootstocks, and avoids the risk of soil erosion. Temperatures in the core of the fire should not exceed 650°C at the most. In very dry conditions, the old vegetation, acting as fuel, will produce volatile and inflammable gases as it heats up in the advancing fire, causing a risk of very rapid and explosive combustion. Burning is a skilled procedure, and good burning systems are generally applied to a mosaic of patches rather than to large unitary tracts. An example of a system of grazing management by burning, the heather moors of Scotland, is described in detail in Box 9.2.

Finally, we must remember that fire is a natural and normal event. To those of us living in the middle latitudes thunderstorms occur quite infrequently. However, in the tropics these are literally everyday events, and lightning strikes cause regular fires. Ecologically these natural fires are important: forests benefit from occasional fires. New vegetation succession processes will invigorate forest vegetation, and some consumer organisms depend on natural fires to give them grazing opportunities not readily found elsewhere. In other words, both forest regeneration and the viability of some animal communities depend on natural burning. Generally humans see fires, other than those used under controlled circumstances for range management, as a bad thing, but as is indicated above, this is not necessarily the case. In areas protected for conservation, the incidence of fire has been much reduced, because fires are reported by vigilant visitors and quickly extinguished. In some National Parks in the USA it has become necessary

Box 9.1

Hedgerows and shelterbelts

Hedgerows are a distinctive and attractive feature of the English landscape. Shelterbelts, lines or blocks of planted tall trees, as well as the lower but similar English hedgerows, are found in many parts of Europe including Britain. Hedgerows are not unique to southern Britain, but are a valued and widespread element in the landscape of this area. The purpose of these features is to act as field boundaries and to provide shelter or shade for grazing animals. This latter is as much a matter of economics as welfare, since protected stock will be in better condition. For all their 'natural' value as landscape components and conservation refuges, these are human artefacts (Appleton 1975). Some hedgerows are medieval, but more are associated with the enclosures of the second agricultural revolution from the eighteenth century onwards. Shelterbelts are even more recent in origin. In Europe many are products of the nineteenth and twentieth centuries, and in other parts of the world almost exclusively so. In the second half of the twentieth century pressure to intensify agricultural production led to the loss of hedges as fields were enlarged to utilise big machinery as fully as possible. This has become something of a *cause célèbre* among those concerned with preserving the beauty of traditional (but man-made) rural landscapes, and countryside conservation.

The main feature of the ecology of these features is that they are composed of plants, which are more akin in ecology to the forests long cleared, though lacking the species diversity or structural complexity of the original. Competitors, rather than **ruderal** species (disturbance-tolerant), dominate the plant community, and animals have a much higher diversity than in intensive farmland. For this reason they have conservation as well as aesthetic value. They provide habitats for natural ('wild') plant species and animals threatened or eliminated in modern intensive agriculture, and thus are useful supports for biodiversity. Hedgerows and to a lesser extent shelterbelts are important in landscape conservation and as amenity woodlands. In the UK small woodlands are often used as preserves for game birds. This raises the question: Are there conflicts between hunting and conservation? In many ways, taking a strictly ecological perspective, there is no difference between controlled game hunting and agriculture, and if wider ends (such as the promotion of biodiversity) can be served, then hunting is ecologically sustainable. There are of course other views on the ethics of hunting of wild animals.

The new directions in European farming, which have followed the reorientation of the Common Agricultural Policy (CAP) in the mid-1990s, have seen landscape quality and conservation dimensions become important in farming policy. In the UK the Environmentally Sensitive Area (ESA) policy has been a way in which planting and protection of small woodlands and hedges has been promoted. The national view is that within the economics of environmental protection and conservation, money formerly directed to subsidising production is now directed to ecological and environmental protection. Although under threat, there now seems to be a more secure future for this important element in the British landscape.

to allow natural fires to act on the vegetation as part of the normal and natural regime of ecological processes. Experiments with artificial burns were both controversial and difficult to control. Fire 'pathologists' are now employed to determine the cause of fires in some of these parks, so that 'natural' fires may be allowed to burn, while those caused

Box 9.2

Heather moorlands and their management by burning

The general pattern of succession has been examined in Box 6.2. How does this natural process relate to human management of moorland? Large areas in the cooler, wetter and hillier parts of northern and western Britain are covered by moorland. Although moorland is found in other parts of Western Europe, the greatest development of this ecosystem is in the British Isles. Moorland ecosystems in most cases are essentially the outcome of human actions. The natural succession is modified and halted to produce a type of vegetation which is primarily for human use. Although the plants involved are not domesticated, and many people regard the landscape as 'wild' and 'natural', the majority of moorlands are to a greater or lesser extent human artefacts. Moorlands are located on hill land with acid soils. Rainfall is variable, but generally there are substantial soil water surpluses for all or most of the year. The growing season is short, six months or less. Under natural conditions, heaths and more competitive acidophilous grasses and sedges would replace colonising bryophytes and vascular plants. At the climax open scrub woodland of oak, pine and birch would develop, depending on local climatic conditions. At higher elevations (about 500 metres in the mountains of central Scotland) climatic conditions are so severe that tree growth would not take place. In these conditions natural climax vegetation would be dominated by heathers such as *Calluna vulgaris*, *Erica tetralix* and *Erica cinerea*, and by grasses such as mat grass (*Nardus stricta*) or flying bent (*Molinia caerulea*), rushes (e.g. *Juncus trifidus*) and sedges (e.g. *Carex bigelowii*), as well as mosses and lichens.

The human role in moorland ecosystem management is to arrest seral development, so that dominant species are heaths and certain grasses that are grazed by sheep, red deer (*Cervus elephus*) and grouse (*Lagopus lagopus scoticus*). Moorland management is practised by systematic burning and by control of grazing. Typically moorland is held by a single owner in large tracts of 500–5,000 hectares, large units by the normal standards of land ownership in Britain. This pattern of landholding is related to the low biological productivity and low economic output of moorland ecosystems. Large parts of upland Britain are managed as estates, in which moors provide the main biological resource. During the late eighteenth and early nineteenth centuries extensive sheep farming replaced peasant semi-subsistence hill farming. In the second half of the nineteenth century, field sports became very popular among the wealthy bourgeoisie. Grouse and red deer were hunted by the growing number of people who had become wealthy as a result of industrialisation in Britain. Large estates were created from the sheep farms, on which land was exclusively or largely managed for the game species mentioned. Often this was combined with sheep farming, but in any case the management system was supported by money earned through industry and commerce, rather than by the actual value of the crop taken from the land. To sustain the maximum amount of plant material available for grazing, the moors were regularly burnt and the numbers of grazing animals were controlled. Thus during a period of over a hundred years the ecosystem has been anthropogenically maintained at a sub-climax stage. Plant species diversity has been reduced and habitat variety diminished as a consequence of this action. Furthermore, soil conditions have been impaired by increased acidification and reduced nutrient availability.

The management of moors by burning shows how deliberate human ecosystem disturbance by burning uses seral change to attain a desired goal. The example that follows deals with the case of heather moors. Heather, the common name for various low-growing shrub species of Ericaceae, forms natural moors or heaths in various parts of Northwest Europe. Heather moors are the natural habitat of grouse, which is an important game bird. Heather moor occurs at low levels on acidic substrate as well as on higher areas. It does not thrive in very wet conditions and thus is best developed in eastern parts of Britain. Grouse shooting, which first became popular in the mid-nineteenth century, is rated to be a high-quality sport; shooting on the best moors is an expensive recreation. The grouse moors of Britain are among the most extensive areas of landscape exclusively managed for hunting anywhere in Europe. The best moors can provide an income from the activity that can not only support the labour required for management, but also provide an important income for the landowner. However, in many cases returns do not match costs of inputs, and either the sport must be subsidised by the owner, or (as is discussed below) the land used for other purposes. Over much of the area occupied by moorland, multiple-purpose use of resources is now normal.

Natural fire is rare in upland Britain, so that climax vegetation, open woodland, is very vulnerable to fire. Tree growth, which is slow in the cold, wet climate, is entirely suppressed by burning. Grazing has the same effect. However, heather species can regenerate quickly after a fire if the rooting system of the plants is not destroyed. Grasses, which are generally disturbance-tolerant, will reproduce quickly and are thus well adapted to repeated burning pressure. Good moorland resource management will attempt to maximise the amount of palatable plant material for grazing. The heather plant can live for several decades. As it grows older it becomes a large, mainly woody plant about a metre in height with procumbent stems. The plant spreads out, and eventually becomes senile and dies out from the centre. In its early stages it forms a short sward composed mainly of green shoots up to 30 cm in height. These shoots provide good grazing, and grouse feed almost exclusively on heather shoots. Burning is carried out about every ten to fifteen years ideally, in rectilinear patches of one or two hectares. A well-managed moor becomes a mosaic of patches of heather at different stages in life cycle, though with little or none at the oldest stages. Some taller heather is needed to provide cover for grouse. The sequence of life stages of heather is shown below. The object of burning is to destroy sub-aerial tissue while allowing roots to survive. To attain this a carefully controlled burn at 400–600°C is required. Too low a temperature of burn will not kill woody tissue, while too high a temperature will kill the plant entirely. In this latter case the whole seral sequence must begin again, much nutrient material is lost by volatilisation and in severe cases soil erosion may result (Gimingham 1972).

The management of moors is particularly interesting because not only does it cause large-scale change to the existing landscape, but it also involves the manipulation of an ecosystem for other than economic reasons. Grouse shooting and deer stalking, though nowadays activities which make a substantial contribution to the income of most estates, started as a fashionable recreation for the rich, and even now, in the vast majority of cases, would be unlikely to provide a sufficient economic return to justify resource management exclusively for this purpose. The heather moorlands so loved by tourists, and thought of as representing the unspoilt natural beauty of Scotland, are little more than managed hunting reserves. However, it is likely that there will be changes in the future. Some sporting estates will survive but others will be unable to maintain the expensive land management needed to provide good shooting. What will happen to moors is uncertain. Some will be converted to intensive sylviculture, but others may become truly wild land again and the arrested seral progression may be allowed to continue to climax. From the conservation perspective the latter is desirable, but generally a wider range of uses and vegetation cover in these areas may be ecologically and economically beneficial. However, it is important that some moor survives. Conservation of heather moors using the traditional burning management regimes is now taking place. Besides the ecological importance of the moorland, moors are attractive landscapes used for amenity purposes by large numbers of people.

by humans are extinguished. A further complication in this somewhat confusing situation is that the majority of fires of human origin are now as a result of arson rather than accident. As this section has shown, the actions of humans may be hard to understand and harder still to manage in ecological systems, so that both conservation objectives are achieved and utilisation of resources is sustainable.

Introduced species

The movement of species from one part of the world to another as a consequence of human actions poses an increasingly serious threat to ecosystems everywhere. These alien or exotic species may be at any trophic level in the system, but generally have enhanced survival rates either because they are more efficient competitors than native species, or because they lack any native predator or disease, or for both reasons. There are a number of reasons why introductions occur. As shown in the following discussion these reasons result in the conclusion that circumstances leading to invasion by an alien species are becoming more common. Thus the threats posed by introduced species are increasing. Although we are aware of this problem it is particularly difficult to control, and once established, introduced species are very difficult to eliminate.

Introduced species may be taken to a new location deliberately. Examples of this include game or decorative species such as pheasant (*Phasianus colchicus*) or rainbow trout (*Oncorhynchus mykiss*) in Britain. The former would be unlikely to survive in large numbers without human assistance, while the latter have thrived and spread from the original points of introduction. Rhododendrons (*Rhododendron ponticum*), brought from the Indian subcontinent to Britain in the eighteenth century, now flourish in hill woodlands to the extent that this species hinders development of natural understorey vegetation and regeneration of woodland. At a more local scale, but in many ways as serious a problem as intercontinental movements, hedgehogs (*Erinaceus europaeus*) have been introduced into the Uists, the southern part of the Outer Hebrides island chain off the northwest coast of Scotland. The controversy surrounding this species and its current status in these islands is interesting. Imported from mainland Britain in 1974 to eat garden slugs, there are now at least 5,000 adults in the islands. Although native to mainland Britain, hedgehogs did not recolonise the Outer Hebrides following the climatic amelioration at the end of the last glacial period, about 12,000 years ago. Egg predation by hedgehogs is thought to be the main reason for the decline in numbers of ground-nesting birds, the conservation of which is of European importance in the Uists. Proposals to remove and/or exterminate the hedgehogs have aroused public passion, both for and against their removal from the islands.

There are numerous examples of accidental introductions, the unwitting effects of which have been profoundly damaging to host ecosystems. The zebra mussel (*Dreissena polymorpha*), a small mollusc originally from the Caspian Sea, has colonised the Great Lakes of North America. These arrived as stowaways in ships' ballast water after the completion of the St Lawrence Seaway in the 1960s allowed ocean-going ships to reach these inland water bodies. Extremely dense colonies of these organisms quickly developed, in the absence of predators. These colonies have caused a significant impact on both lake ecology and human use of water resources. This example demonstrates why introductions are an increasingly serious ecological issue. Over the past hundred years worldwide travel and transport has increased by an order of magnitude, thereby facilitating the spread of organisms. Even remote locations are not immune, as exemplified by the Pacific island of Guam, where snakes, arriving as free-loading air travellers, are now a problem pest, and are causing serious damage to the indigenous ecology. Box 9.3

Box 9.3

The case of the alien fish species Ruffe (*Gymnocephalus cernus*) in Loch Lomond, Scotland

Loch Lomond is the largest lake in Great Britain. As well as possessing great natural beauty it is an important area for biological conservation, and is heavily protected, with National Park status. One of the important species for conservation is a fish, the powan (*Coregonus lavaretus*), which is a relic species linking the loch to its glacial past. It is now found in the UK only in Loch Lomond and in a smaller loch about 20 km to the west, Loch Eck. In the past the abundance of this species of fish, which feeds on zooplankton, meant that it was netted for food up to about a century ago. It itself is preyed upon by larger fish and piscivorous birds such as herons, and is an important element of the loch's food chain.

In the past twenty-five years, however, it has been threatened by a new and alien enemy, the ruffe (*Gymnocephalus cernus*). This small fish species, not recorded in Loch Lomond until the late 1970s, is now the commonest fish in the loch. How the ruffe got into the loch is a story that, with variants, has been repeated in lakes elsewhere in Europe and North America. Ruffe are used for bait in angling, especially for large fish such as pike (*Esox lucius*). This fish was not traditionally a main target for anglers in Scotland. However, with better road communications from the south from the late 1960s, anglers from England could reach Loch Lomond for short weekend visits. Among these were keen pike fishermen. Their enthusiasm together with the lack of a close season for pike fishing in Scotland meant that within a decade pike became rare in the loch. However, the ruffe, which was used as live bait and released at the end of trips, multiplied. Ruffe food includes powan eggs. This predation has had a serious effect upon Powan numbers. There is no practical way of elimination of the alien ruffe. Powan conservation may depend on creation of a new community in an unmodified water body such as one of the reservoirs in the area. The wider potentially serious effects of this introduction on the ecosystem of the loch as a whole is as yet unclear (Adams 1994).

---

considers the case history of an introduced fish and its impacts on the ecology of a large temperate lake.

The fundamental reason why introductions are such an ecological threat is that it is so difficult to return to the situation prevailing before the alien species arrived. This is clearly illustrated by the case of the coypu (*Myocaster coypus*) in the wetlands and rivers of eastern England. These large South American aquatic rodents, which escaped from fur farms several decades ago, reached a population size of 5,000, damaging crops and local vegetation, and undermining banks, before finally being eradicated in 1989 after a campaign which cost over £2 million. Given that there may be no natural local predators for the alien species, introducing predators may seem to be a possible alternative control strategy. However, trying to modify existing ecosystems by further introductions is not always a good idea, and there have been problems with this approach. All of this indicates that it is much better to avoid the problem in the first instance than to deal with the outcomes.

Recreational impacts

Since the 1960s there has been a substantial growth in the use of the countryside for outdoor recreation throughout the developed world. The term 'countryside' includes all kinds of non-urban land from true wilderness to cultivated farmland. Although generally applied to the developed Western world, in this discussion it is used to refer to any part of the land surface of the planet. Recreational use of land and water systems is now growing rapidly in all societies. In many parts of the world, it is recreational use of land and its biological resources that is the dominant rural economic activity, measured both by income generated and by employment. This trend is likely to spread to almost all parts of the world.

The example of countryside recreation in Britain provides a good general model for the examination of the ecological impacts of countryside recreational activities. There has been a remarkable growth in numbers of people involved in countryside recreation since 1960. The causes are the increase in personal mobility, which has come about through increased car ownership, increased leisure time and increased disposable income (Dickinson 1988, 2000). Countryside recreation activities are concentrated in short periods of time and in restricted areas in space. It is this concentration of impacts that is the primary cause of significant damage to ecosystems. A further reason why recreational activity causes impact on the countryside is the nature of the ecosystems which are used for recreation. Recreation often takes place in areas in which ecosystems are fragile and plants are vulnerable to disturbance. Examples of such areas are to be found in the mountain and hill areas in Britain, Europe and North America. These mountain and upland ecosystems are dominated by plant species that are stress-tolerant (see Chapter 5), but are generally much less well adapted to tolerate disturbance. Animal communities, which often include species of conservation importance, are also vulnerable to direct disturbance and impact upon the vegetation cover. Much the same is true of ecosystems in and around rivers and freshwater bodies. In part the attraction of such areas for recreation is related to their wildness or naturalness. Moreover, these areas also provide the resource base for such outdoor recreational activities as hill walking, climbing and skiing, or the wide range of water-based activities which have become popular since the 1960s. When the nature of the ecosystems is taken into consideration with the spatial and temporal concentration of activities that cause impacts, it is inevitable that in the most vulnerable and heavily used locations, outdoor recreation causes serious damage to ecosystems.

Hill walking in Scotland illustrates many of the issues involved in recreational impacts. This type of activity has grown substantially since 1960 (Countryside Commission for Scotland 1992). Hill walking uses mountain paths and tracks which, in the main, were pre-existing agricultural, sporting or forest paths or have been delineated by walkers' use, rather than constructed specifically for recreation. Some of the most popular paths are now largely engineered, a management response to existing pressure problems. Ecological impacts are due to continuing high levels of recreational use during summer weekends. The hill land, through which these trails pass, has anthropogenically modified moorland or mountain ecosystems. Climate is cool and wet, soils acidic, and both species diversity and primary production are low. Scottish moorland ecosystems are dominated by acidophilous grasses or heather (*Calluna vulgaris*; *Erica* spp.) and are sub-climax vegetation communities, in which progression to climax open woodland has been arrested. Mountain ecosystems have vegetation cover dominated by species adapted to the stressed conditions of high elevations. Species include fescue grasses (*Festuca* spp.) and rushes (e.g. *Juncus trifidus*). Such ecosystems are vulnerable to the impacts of outdoor recreation. Recreation is concentrated in time and space; this is reinforced in mountain paths, as the whole of this activity is concentrated on narrow linear tracts which have a small total spatial area.

Depending on the actual level of use, the surface of the path will be stripped of any vegetation cover by the abrasive action of boots. This is exacerbated on steep slopes where shallow cuspidate hollows that have been called 'toe-steps' can form, as is shown in Plate 8a. The underlying substrate is compacted by the load of walkers. Damage is more severe if soils are thin, poorly drained, on steep slopes, or with a low structural consistence in the surface zone. Such conditions are very common in upland areas in Scotland. Removal of vegetation and damage to substrate encourages erosion of the path. Erosion is very largely carried out by surface runoff following the line of the path. The vulnerable precondition is the result of human action, while the actual erosion is an accelerated natural process. Once started, erosion, especially on steep slopes, may form gullies, which may be half a metre or more deep. The process may spread over a wider area as walkers leave the main path to find easier ground on which to travel, thereby widening the affected area. Steeply sloping or rough parts of paths are particularly vulnerable to this type of impact.

Liddle (1975) has characterised the changes that take place in natural vegetation as a result of walker pressure as a kind of reverse succession. Liddle and Scorgie (1980) have also made a similar general review of the ecological impacts of recreation in aquatic environments. Plant species most vulnerable to tissue damage caused by crushing are eliminated first, followed by more resistant species and so on. Eventually the whole surface is unvegetated. The species most vulnerable to crushing are herbaceous flowering plants. Grasses and mosses are more resistant to damage caused by crushing. Pedestrian impact results in compaction of the upper part of the soil profile, thereby impeding surface drainage. Low levels of ecosystem impact can increase the rate of mineralisation of organic matter, which together with the disturbance tolerance of grasses can give a lightly used path an enhanced cover of grasses. However, increased pressures will result in an increase in the area of bare ground and deterioration in drainage. Impacts, which result from horse traffic, trail bikes or off-road vehicles, are similar but act more quickly and severely.

There are two general strategies for path management when damage to the path and surrounding ecosystem happens as a result of recreational use. First, the environment may be modified. This type of management ranges from simple actions such as reseeding and improvements to local drainage, too much more engineered approaches in which steps or boardwalks are used in heavily damaged and vulnerable areas. The second approach is to manage numbers of users. This may be done by restricting access to a particular area, or by creating an alternative route or restricting car-parking. Both approaches may be used together. Footpath erosion may be a problem of restricted spatial dimensions, but in damaged areas the effects are serious, and may extend beyond the line of the path alone. Restoration and continuing management of paths may be a major cost item in countryside recreation. Numerous examples of this problem are found in long-distance trails, mountain tracks and National Parks throughout Europe and North America. Plate 8b shows an example of damage to paths in Scotland.

A way forward – sustainable development

The impacts that have been discussed in this chapter have been chosen to illustrate the range of ecological effects that human actions can produce on ecosystems, and to show that the principles relating to the working of ecosystems, particularly those of functional ecology, are an effective way of understanding human impacts. Furthermore, ecological

Plate 8 Impact on the West Highland Way long-distance footpath, Scotland: (a) looking down-slope, the path has eliminated vegetation cover dominated by the fern bracken (*Pteridium aquilinum*) and common heather (*Calluna vulgaris*). The substrate surface has been exposed. 'Toe-steps' have formed and are linking into small gullies caused by water running down-slope (b), details of the area shown in (a). Bare ground and 'toe-steps' are visible in the middle of the picture

Original photos: G. Dickinson

science is the cornerstone of understanding the scale, extent and severity of problems relating to human impacts on the environment and the biosphere. Our current state of knowledge is far from perfect. While there is a general feeling that human impacts on the biosphere are the most serious issue facing humankind in the future, not all agree (e.g. Lomborg 2001). However, it is most likely that human impacts are a cause of significant changes to the biosphere and at least some of these are threats to the future of humans themselves and all of life on Earth as they both exist in the present.

What, if anything, may be done to address the kinds of ecological problems discussed in this chapter? The answer may lie in the concept of sustainable development. However, this is a difficult concept to define, and harder to use as the basis of resource management strategies. As far back as the period immediately following the First World War, T. Griffith Taylor, a geographer working in Australia, cautioned against what he saw as ecologically and environmentally inappropriate development of the fragile ecosystems of the interior of the continent. These were highly controversial and unpopular views in a country which at that time saw development of its interior as being vital to its future prosperity. Griffith Taylor, by reputation a forthright character who had been meteorologist on Scott's ill-fated Antarctica expedition of 1912, was effectively driven out of Australia. However, he continued his academic work, which focused on the links between human actions and the environment, first in the USA and then in Canada.

In the 1950s, the monumental symposium volume entitled *Man's Role in Changing the Face of the Earth* (Thomas 1956) was one of the first attempts to look at the role of human actions on the biosphere. Set in a time of post-war optimism, when no limits to economic growth could be seen, the contributions by some of the most influential scholars of the day sound voices of concern time and again about the scale of human impacts on the world.

These are two examples of the early re-evaluation of the impact of humans on the planet. There are many more, particularly in the past forty years. In the past four decades, the growth of what is now called *environmentalism* has gone forward to the point that not only is its agenda a part of mainstream politics, but also it increasingly dominates that agenda. Environmentalism is an ethos, which makes biological conservation and sustainable use of the biosphere and the abiotic environment the guiding principle for human activities. The historical development of environmentalism and conservation is summarised by Primack (2002) from an American perspective, and by Pullin (2002) from a British viewpoint.

A critical event developing the idea of sustainability was the Brundtland Report, named after its chairperson Dr Grö Harlem Brundtland, Prime Minister of Norway. The report from her group, the World Commission on Environment and Development entitled *Our Common Future* (WCED 1987) was commissioned by the United Nations to provide an agenda for change to meet the needs of all people, to use the planet's resources wisely and to build these into national and international policies of all nations, rich and poor. The issues that the Commission saw as central were the human population, its supply of food, loss of biodiversity (though the Commission did not use this term) and the human factors of energy needs, industrialisation and settlement. These, as the Commission pointed out, are all closely interconnected. The key to resolving the tensions between human society and its sustaining environment and biosphere was to be sustainable development. In an elegant and frequently quoted definition this was 'development that meets the needs of the present without compromising the ability of future generations to meet their own needs'. However, the Commission also reported that sustainable development must both address the needs of the poor and operate within the limits set by human societies and their technologies to meet future needs. A wider, critical review of sustainability is given by Mitchell (2002).

A major political landmark in fixing the idea of sustainable development was the Rio de Janeiro 'Earth Summit' held in that city in 1992. Mitchell (2002) gives a good description of this titanic event. Three main outcomes were agreed by the world leaders attending, which related to forest resources, climate change and sustainable development. Although the success in putting these into action has been at best patchy, this was a huge step forward in placing environmental concerns and good stewardship of the environment at the heart of the global political agenda. The main difficulty was, and is, that achieving outcomes depends upon people's attitudes and aspirations and experiences. Linking these to sustainable development is the challenge for the future.

This book is concerned with ecosystems rather than with sustainable development. Therefore it is in terms of ecosystem functioning, and the ways in which this is fundamental to sustainability, that we examine the issues raised above. Martinez (in Gaston 1996: 114–48) reviews the ways in which biodiversity relates to ecosystem functioning. The key issue in linking biodiversity to ecosystems is that the functioning of ecosystems should lie within the parameters of change that occur as a result of patterns of normal variation, bearing in mind that ecosystems often follow non-linear behavioural trajectories (i.e. they can sometimes change rapidly and rather unpredictably). This link between ecosystem functioning and biodiversity, the core concept in contemporary conservation, is a measure of the importance and continuing relevance of the ecosystem concept and of the functional ecology approach.

In the following two chapters we look at the large-scale impacts of human activities and their ecological consequences for ecosystem functioning. In these chapters we show the value of a functional ecology approach to ecosystems, both in theory and examples of human impacts. In particular it is through the systems approach employed in ecosystems that we have a means of understanding the complexity of human impacts on ecosystems. The ecosystem approach provides us with a powerful tool to deal with the environmental issues that result from human impacts on the biosphere.

Summary

- This chapter considers the nature of interactions between humans and ecosystems. Although humans may be considered as biological parts of ecosystems, the effects of human activities, which have come about as a result of the rapid increase in population and the consequences of industrialisation and intensive agriculture, are so profound that it is appropriate to consider human impacts on ecosystems separately from normal ecosystem function.
- Impacts on ecosystems are varied. Some of the most widespread and serious impacts are considered through a number of examples.
- The concept of sustainable development is introduced as a strategy for minimising human damage to ecosystem function.

Discussion questions

1   Identify the impacts upon ecosystems which might occur as a result of the construction of (a) a plant manufacturing semi-conductors in a redeveloped riverside location; (b) an office block for an insurance company, adjacent to the green belt of a large city.
    Do this for a location in Western Europe or eastern USA and then for a location in southern India or southeastern Brazil. What differences in impact on ecosystems would be expected, and what human dimensions of impact would be different?

2   Think of three general arguments that you, from the developed world, could put to someone in government in a developing country, as to why conservation of an endangered species is

important. Review the ways in which this might affect economic development within the developing country.

3   Is biological conservation the same as protection of natural ecosystems? If not, under what circumstances would conservation mean something else, and what actions would be involved in this sort of conservation?

Further Reading

See also

Disturbed ecosystems, Chapter 6
Large-scale impacts on ecosystems, Chapter 10
Global environmental change and consequences for ecosystems, Chapter 11

Further reading in Routledge Introductions to Environment Series

Natural Environmental Change
Environmental Biology
Energy, Society and Environment

General further reading

Changing the Face of the Earth (2nd edn). I.G. Simmons. 1996. Blackwell, Oxford.

An eloquently written analysis of how humans have affected the Earth, and an evaluation of current environmental and ecological problems.

Environmental Issues in the 1990s. A.M. Mannion and S.R. Bowlby (eds). 1992. Wiley, Chichester.

Most of the chapters in Part 3 of this book are highly relevant. It is also most useful for Chapter 11, because it examines impacts at global level.

The Human Impact on the Natural Environment (5th edn). A. Goudie. 2000. Blackwell, Oxford.

A comprehensive analysis of human impact on natural systems that is up to date and well written.

The Human Impact Reader. A. Goudie (ed.). 1997. Blackwell, Oxford.

A collection of research studies related to the impact of humans on natural systems. Part 5 on biological impacts is most relevant, but all the chapters in this book are helpful to the appreciation of human impacts on ecosystems.

10 Large-scale human impacts on ecosystems

This chapter deals with larger scale and widespread impacts on ecosystems, focusing on four examples. As was seen in Chapter 9, human impacts are not new. There is good historical and archaeological evidence that human impacts on ecosystems have been serious in the past. However, both the scale and extent of impacts have increased rapidly in the past century, and they are not confined to either the developed or developing worlds. The issues we discuss here are global and increasing in their extent. Nevertheless, there are ways in which the effects of these impacts can be significantly reduced. Solutions involve difficult economic and political issues. If humankind does not address these widespread impacts on ecosystems resulting from the actions discussed in this chapter, the consequences may be as serious as that of global climatic change, which is considered in the final chapter. In general, problems associated with ecological impacts are interrelated, and their solution must involve all people, including those not directly affected by the impact. This chapter covers:

- **Soil erosion**
- **Eutrophication and nitrate pollution**
- **Desertification**
- **Deforestation**

The characteristics of large-scale human impacts on ecosystem function

In Chapter 9 we analysed examples of human impacts on ecosystems, looking in particular at impacts that influenced the trophic structure of ecosystems. In this chapter we turn our attention to large-scale impacts on ecosystems. These may be defined as impacts that have effects on all or most parts of the ecosystem, including its abiotic environment and general functioning. They are to be found throughout the biosphere; that is, everywhere humans are or have been. These large-scale impacts have four general characteristics:

1   Reduction in biodiversity
2   Altered gross primary productivity, and energy flow
3   Gross change to nutrient cycling systems
4   Difficulty in reversing the effects of impacts, through reduction in ecosystem resistance and resilience

These characteristics relate closely to the strategy of ecosystem development proposed by E.P. Odum (1983: 444–68). Odum developed these ideas in the context of the

Table 10.1 Biodiversity and ecosystem functioning

Species role characteristic	Biodiversity trajectory	Biodiversity outcome	Importance in ecosystem function
Redundancy	Insensitive – flat trend in relationship to variation in biodiversity	Neutral – does not add to total biodiversity as redundant species is replaced by another	Specific but limited role in both time and space
Singularity	Sensitive – may result in either a positive or negative trend	Makes a unique contribution to biodiversity. Irreplaceable	Addition or loss causes change in functioning. May well be a **keystone species**
Idiosyncracy/ unpredictability	Variable in time and space	Variable, though is often locally significant in biodiversity	Typically variable according to specific site and environment

theory of autogenic succession (ibid.: 446), but the trends can equally be described using the concepts inherent in C-S-R strategy theory. Human impacts tend to reduce the role of competitors, which are less able to cope with the disturbed environments produced by human actions. This reduces primary productivity and may affect biodiversity if a more restricted group of disturbance-tolerant species becomes predominant. If the human impact is sufficiently powerful, ultimately stress-tolerant species may become the major group of plants present. Plant communities dominated by species in these latter two categories are generally inherently less productive than communities dominated by competitors (that is, climax communities, as discussed in Chapter 1) in the same climatic environment. However, this view of ecosystem strategy is by no means universally accepted. Many ecologists contend that the end-point of succession occurs when nearly all community respiration is needed to maintain community biomass (i.e. there is a balance between energy production and consumption within a community: Colinvaux 1993). From this perspective, succession is regarded as the outcome of 'individuals of species with different properties acting in ways that maximise individual fitness' (ibid.: 441).

A more recent development has been the statement of hypothetical relationships between biodiversity and ecosystem functioning. (Naeem *et al.* in Loreau *et al.* 2002). The basics of this theory are shown in Table 10.1.

In the following ecological analysis of large-scale impacts our arguments generally support the theories of succession put forward by Odum to explain biodiversity and ecosystem functioning. Thus we contend that the ecosystem concept and a functional ecological approach to ecosystem analysis are valid at the large as well as the small scale.

Theories about change in ecosystems and the environment

Environmental change, whether natural or human-induced, operates in very complex ways. Changes rarely act consistently in one direction, or at the same rate for long periods. This means that it is difficult to predict how the environment will change in the future, even when good data about existing and past environmental circumstances are

---

Box 10.1

Davisian cycle: an explanatory and critical commentary

The Davisian cycle was developed by one of the founding fathers of modern geo-morphology, the American William M. Davis (1850–1934). It was the most influential theory in early geomorphology, though its current status is more of a historical curiosity. Indeed some contemporary geomorphologists complain of the lasting strait-jacket of Davisian thinking.

The model proposed that the evolution of landforms over time led to the develop-ment of characteristic types of landforms. The development of landforms in an area would pass through a series of stages that were millions of years in length. Initial stages involved the dissection of new mountain or upland areas, following mountain building or uplift. As dissection proceeded, relief would become more subdued, until in the final stage a peneplain or area of almost complete lack of relief remained. Subsequent uplift would start the cycle again. Davis's theory was supported by his own observations, but this has been (legitimately) criticised as being subjective, and at variance with objective measurements and analysis. Davis suggested that this was a cyclical system, operating over millions of years to pro-duce replicate landform systems according to a particular stage in development within the cycle (Goudie 1984).

This model was highly influential in the development of the science of geo-morphology. However, since the 1960s this view has been challenged, and most geomorphologists now reject this type of model which is seen as not fitting the geomorphologic evidence and being replaced by better theories. In general, eco-logical and environmental science models which project development towards an end point or equilibrium 'goal' have been criticised as being simply deterministic and not fitting research evidence. Much recent research supports the view that most environmental and ecological systems operate in a non-linear manner, with stochastic components in their system behaviour. The outcome of such processes is much more difficult to predict. Outcomes are various starting from the same ini-tial set of conditions.

For more details about the Davisian theory and some alternative views see Goudie (1984: 241–4).

---

available. In the first chapter of this book, we saw how much of the pioneering research work in environmental and ecological science led to the development of models of change over time. Clementsian succession, as discussed above, is a good example of such a model. The Davisian cycle (Box 10.1), which was also developed at the beginning of the twentieth century, is an example of a model of systems behaviour in the abiotic environment. Davisian theory was very influential initially but was criticised by later workers. Better measurement of ecological and environmental systems in particular cast doubt on the widespread validity of this type of theory. Furthermore, these are examples of theories which do not include human impacts as components in the system. Theories that can explain and predict the relationships between ecosystems and the changing environment must include forcing factors of human origin.

As ecological science progresses, for example, by incorporating non-linear dynamic approaches (based on the mathematics of chaos theory), better models of the precise functioning of ecosystems will undoubtedly be developed and tested by empirical research (e.g. Schroder *et al.* 2005). However, even with these refinements the eco-system is likely to continue to provide a useful framework for the investigation of inter-actions between the living world and its abiotic environment, and one in which the impact of human actions on the biosphere can be identified and analysed. Without an integrative framework, the true nature of environmental change, human impacts and the threat to the functioning of the biosphere and our life support systems cannot be understood.

Soil erosion

Soil erosion is one of the most serious problems caused by human activity, and one which has occurred all over the world throughout historical time. Removal of the par-ticles that make up the physical soil framework by the natural agencies of water and wind is a normal and continuous process in all kinds of environments. However, human actions, such as deforestation, or agriculture practised at levels of production beyond the **carrying capacity** of the resource base, are likely to lead to removal of sediment at a rate several times faster than by natural processes. The cause of this accelerated erosion is change in the balance between energy available to transport material and the available supply of material, sediment, for transport. Natural processes cannot reverse the effects of such impacts readily or, in some cases, at all. Soil once eroded is unable to support a normal cover of vegetation, and is reduced to a condition in which normal ecosystem functioning cannot take place. The damage caused by soil erosion often takes decades to repair, and the restored soil may never return to its original state. Significant and large-scale damage to natural systems, and thus reduction in the resource value of such areas, has occurred since the times of classical civilisations in the Mediterranean (Simmons 1996). Effects may be exacerbated by climatic fluctuations and are often driven by the pressure of population on resources.

The examples of the 'Dust Bowl' in the Great Plains states of the USA in the 1930s, and of the Sahel in Africa during the 1970s and 1980s, show that such disasters can happen in both the developed and the developing worlds in recent times (Myers 1985; Cloudsley-Thompson 1989; Mannion 1991). Lest it is thought that soil erosion is now exclusively a problem of the Third World, in 1992 more than 60 per cent of US crop-land was experiencing soil loss (Cutter and Renwick 1999), and this situation has not improved since then. It is notable, too, that the erosion problem is manifest in all environ-mental regions of the country. In advanced agricultural systems soil erosion is the result of a number of actions. Long-term monoculture made possible by the use of herbicides and pesticides to control weeds and pests, together with obligatory massive inputs of synthetic fertilisers, has greatly reduced soil organic matter content. This affects soil ecosystem function, and leads to impaired formation of soil aggregates called **peds**. Soil organic matter content in many of the most important agricultural areas of Europe and the Americas has fallen considerably over the past fifty years (Gray 2004: 155). A further problem is the damage done to soil by heavy machinery, the use of which has increased greatly over the same time period. In much of the developed world intensive agriculture has so reduced the ability of soil ecosystems to function normally that the agricultural output from these areas is almost totally dependent on human actions to modify soil properties. To a worrying extent this is not sustainable.

Eutrophication and nitrate pollution

The issue of eutrophication has been mentioned already several times in this book. It is a worldwide issue. It affects aquatic ecosystems, and because of the physical mobility of aquatic systems the cause and effects of the problem spread quickly. Eutrophication is caused by the boost in primary biological production that results from increasing the nutrient supply in water bodies. It has the most dramatic effects on freshwater bodies in which a relatively small increase in nutrient supply may boost primary production very considerably. In many freshwater bodies phosphorus is the limiting factor. Phosphorus levels may be increased from two main sources. Synthetic fertilisers are generally rich in phosphorus, which being available only to plants in the anionic form of phosphate ions, cannot adhere to soil colloids which also have a negative charge, and are thus easily leached out of soil by drainage water. The second source of phosphates is domestic sewage and waste water discharges. This material is particularly rich in phosphates. Household detergents in particular are a major source. A difference between these two sources is that agriculture is a diffuse source with input of the pollutant coming from an extensive area, while domestic pollution is a point source coming from an identifiable and specific location. In general it is easier to remedy point source than diffuse source pollution. The latter may be rectified only by control of input of polluting material to drainage systems, while the former may be treated at input or discharge. In rural areas farming is an obvious agency of eutrophication, but increasing settlement is also a potential problem if good waste water management is not in place. Settlement, either for commuter houses or second homes, has spread into the remoter countryside throughout the more economically developed world in the past half century. This issue is now spreading to the less economically developed world, where it may be more serious due to lack of legislative and practical controls on discharges.

At worst, eutrophication will cause near complete loss of life in water as the resultant algal blooms deoxygenate water. The result is severe damage to both aquatic ecosystems and water resources. Examples of eutrophication issues include European lakeside holiday homes using traditional existing dwellings, which often do not have mains sewerage systems, resulting in discharge directly or indirectly to the receiving lake. Such problems are now addressed vigorously, not least because water quality standards are expected to rise with the adoption of European Water Framework Directive (WFD) standards for water quality from 2007 onwards. In tropical areas increase in the nutrient content of waters may result in the uncontrolled spread of macrophytic weeds such as water hyacinth (*Eichhornia* spp.) (Pieterse and Murphy 1993). These cause enormous problems, not least when they are also introduced species. The costs and labour associated with control of weeds are very great.

In most cases where agricultural activities are the source of input of nutrients into ground water and water bodies the main problem is with nitrogen. Nitrogen is the inorganic nutrient required by plants in greatest absolute amount (see Chapter 4) and so enters water through leaching in largest quantity. Nitrogen, in the available nitrate form, is anionic, thus not adsorbed by soil colloids, and is easily removed from the soil in drainage water. Contamination of ground water by nitrates is a serious problem in Europe and the USA. In the latter country it is the main cause of closure of public water supplies, and is especially serious in areas where there is a surplus of rainfall over evapotranspiration, areas in which water is drawn from shallow wells from ground water and areas in which there are high-value crops, the growth of which is boosted by heavy use of nitrogenous fertiliser. This problem is difficult and costly to treat. Conventional filtering does not affect nitrate pollution, which must be addressed through reverse osmosis or ion exchange.

The nitrates issue also illustrates the problems which water pollution can pose to human health. Relatively little is known about long-term exposure to low concentrations of potentially harmful materials in the environment. The toxicity of a substance is often measured by LC_{50}, the lethal concentration of the material that will kill 50 per cent of a population. There are a number of issues that make this measure difficult to apply in the real world. First, it is generally assessed through trials on laboratory animals, not humans, and thus is at best an estimate. Second, when used as that simple value, it does not indicate how sensitive individuals may react adversely to low levels of the substance. For this reason, generally much lower dosages or exposures are used in the estimation of safety limits. However, these safety levels are for a single event, and do not relate to risks of long-term low-level exposure. For example, a contaminant risk related to drinking-water, such as nitrates, may be based on daily exposure for decades. The level of nitrate permitted in drinking-water in both Europe and the USA is now very low, well below toxic thresholds. A maximum value of 10 ppm of nitrate-N is allowed in the USA, and most drinking water is actually significantly less contaminated. In Europe the EU Water Framework Directive (WFD) with such policies as the Nitrate Directive (91/676/EEC), under which land areas contributing to nitrate pollution are designated Nitrate Vulnerable Zones (NVZs), address this issue. One final concern is that nitrates, as for all pollutants, may act in combination with other substances to result in an increased hazard. There is some evidence that nitrates react with small amounts of arsenic in this synergistic manner.

Desertification

The term 'desertification' (Table 10.2) is credited to Aubréville, a forest ecologist in 1949. Like a number of key ecological environmental concepts, its definition has evolved over time. A recent widely accepted definitive statement from the United Nations Environment Programme is 'land degradation in arid, semi-arid and dry subhumid areas resulting mainly from adverse human impacts' (Tolba and El-Kholy 1992: 134). This definition stresses the importance of human actions as the cause of desertification, a perspective with which we agree. However, some views of desertification give natural environmental change a significant causal role, and there remains a substantial degree of controversy about the exact nature of desertification. These debates are examined by Binns (1990), Thomas and Middleton (1994) and Warren and Agnew (1988). Whatever the exact current situation, it is certain that human-induced global climate change, the ecological effects of which are discussed in Chapter 11, will increase desertification problems. Desertification is sometimes seen as a problem of the less economically developed intertropical world. Although the manifestations and consequences of desertification are most severe in these parts of the world, it is a global issue, affecting all continents except Antarctica.

The general causal factors of desertification promote a range of ecological and environmental processes which are often interlinked, and in turn will lead to socio-economic consequences. In less economically developed countries, for example, the nations of the Sahel, the arid and semi-arid area immediately to the south of the Sahara in West Africa, sheer economic pressure, which may be exacerbated by population growth, leads to poor resource management. This is not a deliberate preference for people in these countries but is forced on them as a matter of survival. Furthermore, amelioration and adoption of better, ecologically sustainable resorce management may be blocked by inefficient and corrupt economic systems, or wars and civil strife. Poverty and famine have both environ-

Table 10.2 The characteristics of desertification

Causal factor	Expression in ecological and environmental processes	Resultant socio-economic effects
Overgrazing – livestock numbers exceed rangeland carrying capacity	Loss of ground cover of vegetation; soil erosion and sedimentation; deterioration of pasture as a result of overgrazing accelerates processes	Reduced livestock productivity as pasturage decreases and resultant spiral of falling income and food output
Inappropriate cultivation – cropping on land with insufficient raifall to sustain crops without irrigation	Soil erosion and sedimentation, especially as soil organic matter is lost	Crop failure; in extreme cases may result in famine as possibilities for future cropping are lost
Deforestation – removal of wood, especially in semi-arid areas for fuel wood, or as result of browsing pressure from domestic animals	Soil erosion and permanent damage to soil resource damaging whole ecosystem function; regrowth of trees may be impossible	As fuel wood remains a major issue in some less economically developed countries, search for new sources extends the affected area and increases the cost of fuel for cooking
Salinisation of soils – a result of poorly managed irrigation	Soils become uncultivatable due to high pH/alkalinity and waterlogging; may be almost impossible to restore affected areas	Loss of land with a high level of crop output; waste of scarce capital resources, as investment is lost
Climatic varaiability – particularly variations in rainfall	Extent of desert margin varies varies over time (years to centuries)	May cause local problems but in many has been accommodated by indigenous resource management systems

mental and human dimensions, but the lessons of desertification are that the causes of the problem are fundamentally human, and their solution requires good governance as well as ecological knowledge.

The main human causes of desertification, a potentially catastrophic large-scale impact on ecosystems, are deforestation (which is discussed more fully in the next section) and overgrazing. Overgrazing is a problem which affects **rangelands** all over the world. As has been noted, overgrazing is often associated with deforestation in the developing world. However, few parts of the world have not suffered from the problem in historic times. In common with other impact problems, it is not simply a result of human greed and ignorance, though in some cases these may be factors in the equation. The fundamental cause is more frequently economic necessity, often driven by population pressure. Although traditional pastoral societies may have built up good indigenous knowledge of their resource base and its sustainable use capacity, externally induced pressures may force this to be ignored. In some cases there are considerable gaps in understanding the functioning of grazing land ecosystems. In many instances conservation of endangered species and their habitats may have a low priority. Overgrazing damages a wide range of ecosystem properties, and is not simply confined to primary producer species. The primary impact is on the plant community, but changes therein will affect higher trophic levels, decomposers, the soil system and the physical environment of the whole ecosystyem. Although some ecosystems exploited for grazing have a high degree of resilience to grazing pressure, others are inherently fragile. Grassland ecosystems in semi-arid and arid environments are especially vulnerable. Generally, overgrazing is likely to cause permanent change to the affected ecosystem. In some instances the resulting damage to the ecosystem may be catastrophic.

The general ecosystem consequences of overgrazing are reduced production and degraded biodiversity. The sequence of events is usually as follows. High grazing pressure results initially in decreasing producer biomass and vegetation cover, as plant tissue is consumed faster than it is replaced by new growth. This affects species in a differential manner as the plant species most palatable to grazers are consumed preferentially. If such species are major components of the plant community, which is the case in most heavily used grazing lands, the impact will be substantial. Continued pressure will result in the appearance, and spread, of unvegetated areas, and loss of habitat variety, and sometimes invasion by unpalatable weed species. At this point species of high conservation priority may become endangered or lost. It is not only the natural ecosystem which is damaged. Loss of biomass and species diminishes the resource value of an overgrazed area. This in turn increases overgrazing in an exponential manner unless prompt preventive action is undertaken. This must include lessening of grazing pressure, and protection for threatened plant and animal species. Left unchecked, overgrazing will progress to severe ecosystem degradation and soil erosion. Erosion occurs because the unvegetated soil surface is vulnerable to wind and water action, and because reduction in dead organic matter contribution to the soil reduces humus content. This in turn leads to lower structural stability. Much of the desertification which has taken place in the Sahel, to the south of the great Sahara Desert, has been caused by overgrazing. This was driven by population pressure, and compounded by unstable and poverty-ridden human societies, a fluctuating semi-arid environment and poor knowledge of the changed resource base and its use. Similar problems occur in South America (see Case Study 6).

The sustainable solution to the problems of desertification worldwide is to recognise the underlying social and economic problems which cause people to try to keep too many animals on inappropriate land, and to neglect its management. In Argentina these causes are likely to be very different from those pressures which apply in the Sahel. However, if the root causes can be successfully identified (whether they are, for example, problems of over-population and food shortage in West Africa, or absentee ownership coupled with low prices for beef in South America) and people's attitudes altered, then the practical solutions are usually the same everywhere. The introduction of effective rangeland management (i.e. management of the cattle herds, by fencing or active herding, with a reduction in herd size per unit area of land), coupled with controlled burning, has been shown in Argentina to be rapidly effective in restoring the sustainable productivity of desertified shrubland ecosystems (Busso *et al.* 1993). An example of the potential use of resources in a desert area that has been modified by a new water source is shown in Box 10.2.

Case study 6

An example of desertification has occurred in the extensive Caldenal shrubland region of semi-arid southern Argentina. Here the combination of a too-high stocking density of cattle and an inappropriate burning regime has led to serious desertification problems. Fire is a normal feature of dry rangeland ecosystems prone to summer thunderstorms and consequent lightning strikes. Shrub communities in this type of rangeland ecosystem are largely fireproof, either because the plants have their growing points below ground where they are safe from the flames, as in the case of the protected meristems of the native grass species of the area, or, as in the case of the larger trees and shrubs (such as *Prosopis caldenia*) by virtue of a thick fireproof bark. If regular burning does not take place then the quantity of dry litter and dead wood piles up on the soil surface to the point where, when a fire does happen, the temperature reached is intense – too high for the plants to survive. Controlled, regular burning is thus a necessary feature of a well-managed shrub rangeland. In the Caldenal a normal stock density for cattle is in the order of 1 cow per 10 ha. All too often the cattle are kept at higher densities. When this occurs, and if at the same time burning has been neglected, the result can be catastrophic. The palatable and more nutritious grasses and shrubs are destroyed by overgrazing. Very hot fires burn large holes in the remaining plant cover (Busso *et al.* 1993), leaving the soil open to wind erosion by the strong winds which sweep across the plains of southern Argentina. Invasive non-palatable weedy shrubs (such as *Geoffrea decorticans*) rapidly colonise the bare areas; the net effect is to reduce the carrying capacity of the land for cattle to half or less of what it should be. Alongside this economic consequence of the desertification process there is a heavy price to pay in terms of damage to the biodiversity support function of the shrubland ecosystem. A significant proportion of the very diverse bird fauna of Argentina is represented in the Caldenal ecosystem: damage to the vegetation inevitably has undesirable consequences for the habitat and long-term survival of these bird species in South America.

Box 10.2

Environmental and ecological changes in the Wadi Allaqi area of south-eastern Egypt

The Allaqi area of southern Egypt illustrates the ecological effects of the construction of a large reservoir. The location of this area is shown in Figure 10.1. Wadi Allaqi is the largest east bank wadi running into the Nile in southern Egypt. A wadi is a valley that has been formed largely by fluvial action, but in which there is currently no surface water. The valley was formed at times when rainfall in the area was higher than the present. There are occasional rain storms in the Red Sea Hills to the east, but surface flow happens no more commonly than for a day or two each decade. There is, however, underground movement of water down the wadi to the Nile. The wadi is flat floored and in its lower stage has a very low gradient. The hot hyper-arid climate permits only the most drought resistant plant community to evolve, which is reliant on either sub-surface

Figure 10.1 Wadi Allaqi area of southern Egypt

water in the wadi, or water from the rare rain storms. A distinctive and fragile stress-tolerant plant community has evolved in this area.

Following the construction of the Aswan High Dam, the reservoir created, Lake Nasser, flooded the lower wadi. This changed environmental conditions greatly in the area along the new shoreline. Plant species such as *Tamarix nilotica*, found along the banks of the Nile, have invaded the area, and there was a great increase in total biomass. The level of Lake Nasser varies continuously. This is in part due to the reservoir being replenished each year by the flood surge of the White Nile, and in part due to abstraction of water from the lake for irrigation. Furthermore the maximum level of the lake varies from year to year, depending on the amount of water entering the reservoir. This is highly variable. The typical annual variation in the height of water in the lake is 6 m, and since 1978 the absolute variation in lake level has been more than 25 m.

This means that the shoreline and flooded part of the wadi changes considerably. The flatness of the profile of the wadi means that since 1978, more than 25 km of the wadi has been affected by flooding. The altered ecological environment of the shoreline has affected a substantial zone, which is now colonised by a new plant community. This represents both a potential new resource base, and the loss of other resources. Research in this area has attempted to analyse the nature of environmental and ecological changes, and to evaluate how these may be used in sustainable development. The whole area is now protected by the Egyptian Environmental Affairs Agency, and is a UNESCO Man and Biosphere programme reserve. Resource management of the area is directed at protection of remaining hyper-arid biota of conservation priority, and at developing grazing activities, and use of natural vegetation for fuel and medicines.

For more information see Pulford *et al.* (1992) and Dickinson *et al.* (1994).

Deforestation

Deforestation, the loss of forest cover, is a process that has been carried out as the direct and indirect result of human activities since the human species evolved. It is not that trees are enemies of humans: quite the opposite, for trees are one of the most valuable biological resources. It is that trees get in the way of other human activities and deliberately or accidentally are removed. Deforestation is the replacement of natural forest areas by other types of ecosystems, principally agricultural ecosystems. Deforestation is as old as *Homo sapiens*. Primitive humans used fire to clear forest areas for grazing and croplands. Some types of grassland, such as savannah, are almost certainly partly anthropogenic in origin. Most tree species are less able to cope with repeated cycles of burning than grasses, especially if fires occur at relatively short intervals. Grazing also suppresses tree growth as the seedling stage is vulnerable to grazing, and the life cycle of trees is much longer than that of grasses. The **intercalary meristems** of grasses give them great competitive ability when subject to grazing pressure. Deforestation is frequently linked to overgrazing, though even controlled grazing means the loss of tree cover. In recent times deforestation has continued apace throughout the world. As well as clearance of woodland for agriculture, deforestation has been caused by demand for wood for fuel, pulp and construction materials.

The demand for all wood products increased considerably during the twentieth century. A significant part of the supply of soft wood products which are derived from coniferous trees is now supplied by planted forest. This has been aptly described as tree farming. Although it does not destroy natural woodland it is the cause of a number of ecological problems, such as soil acidification. Loss of natural forest ecosystems always results in loss of biodiversity and often in a range of environmental impacts. Demand for tropical hard woods such as teak, mainly from the developed world, is still largely met from natural or semi-natural forests which are mainly located in the developing world. In many cases extraction of a small number of economically valuable individual trees from a forest area leads to destruction of the whole forest ecosystem. Tropical forests are also used by indigenous people. In most parts of Africa, wood is an important source of fuel for cooking. It is unlikely that this demand will be replaced by other fuel sources in the short term. Therefore as population growth continues, deforestation will accelerate. Already around many large cities in Africa there are deforested zones tens of kilometres in diameter. In Zimbabwe, planting of new forest around the capital and largest city, Harare has been carried out to supply future demand. Although economically sound and of some ecological value, the planted forest is of lower species diversity and lower ecological value than the cleared natural forest.

The ecological impacts of deforestation are numerous. Biodiversity in tropical forests is among the highest of any ecosystem. Destruction of forest and its replacement by secondary forest or by agricultural systems inevitably leads to species extinction. The exponential increase in the rate of losses of species during the twentieth century is to a considerable extent the result of the destruction of tropical forests. One estimate for extinction indicates that some 4000 species of mammals and 250,000 species of flowering plants, representing 4.0 per cent and 0.2 per cent of the respective taxa, have become extinct since 1600 (Primack 2002), placing the current mass extinction event in the same league as previous events such as the elimination of nearly all species of the dinosaurs during the Triassic geological period (see Chapter 1). Many ecologists believe that the loss of species will not only be disastrous for tropical forest ecosystems, but also damage overall biosphere functioning. Forests, especially tropical forests, are the greatest store of all types of biodiversity in the biosphere. Furthermore it is becoming clear that, just as advances in biotechnology are beginning to open up new ways in which the renewable biological resources of the forests may be utilised, human destruction of these same forests means that some of that potential resource base is being lost for ever. The complexity of the ecosystems in the humid tropics means that an extraordinary diversity of life is found at all trophic levels. Species loss is by no means confined to primary producers; species in the higher trophic levels and the decomposer chain also disappear.

Deforestation has a serious effect on the physical environment. Removal of tree cover will increase the rate of soil erosion, often dramatically. A considerable part of the energy of impacting raindrops is absorbed by tree canopies. In the tropics where the kinetic energy of rain splash is high, the result of deforestation is the initiation of erosion by displacement of surface particles. Loss of dead organic litter leaves the soil surface unprotected, and structural aggregation impaired. Surface runoff is accelerated, and fluvial action will quickly strip away sediment from the soil surface. An increased flood risk in rivers is a further consequence of deforestation. The rise in occurrence of severely damaging floods in Bangladesh has been linked to deforestation in the upper courses of the great rivers which flow through that country into the Bay of Bengal (Mannion 1991: 250–4). Increased soil erosion and flood risk have meant a disruption in the sediment budgets of deforested catchments (Grainger 1993). Human actions to control floods and erosion require reforestation. Without this vital action, flood control dams will silt up, and soil erosion is likely to continue. In drier areas, erosion may be caused by the action of winds. Removal of vegetation cover in general, and trees in particular, increases the rate of this aeolian action appreciably. Large-scale deforestation may cause changes in local and regional climate. Conversely, human-induced climatic change is likely to have a considerable impact on the nature and extent of forest ecosystems throughout the world. The former issue is well illustrated by reference to the Amazonian rainforest. Deforestation in this, the largest remaining area of tropical rainforest biome in the world (see Chapter 8), will lead not only to soil erosion and loss of biodiversity, but also to regional-scale climatic change. Change in **albedo** is likely to increase surface temperature. Most of the moisture entering the atmosphere in this region is contributed by transpiration. Removal of tree cover is almost certain to result in a drier climate, since transpiration by trees is greater than that by other types of vegetation. This in turn will reinforce deforestation, making the regional climate in the remaining areas of the natural forest drier, which will hinder natural regeneration of the rainforest (Roberts 1994; Whitmore 1998). This will reduce the economic and ecological resource value of the ecosystem. As is discussed in Chapter 11, human-induced global climate change will also affect tropical rainforests. These effects will, of course, be additional to losses due to direct human deforestation.

Loss of the rainforest of Amazonia is the best-known, the most serious and in some ways the most contentious example of these problems. Many ecologists are of the view that, of all the areas in the world where human impacts are damaging natural ecosystems, the consequences of damage to this, the largest remaining area of unmodified tropical rainforest in the world, will be most critical, both for human life and for the biosphere as a whole. Box 10.3 explains and evaluates the Amazonian problem and its human significance. However, the problem is not confined to Amazonia, to Brazil or South America. It is a pan-tropical, indeed a global issue. All humans are affected, and all must play a part in saving forests.

Box 10.3

The problem of forest clearance in Amazonia: an evaluation of the issues

It is widely believed by ecologists and environmental scientists that destruction of rainforest in the Amazon basin is one of the most serious impacts upon ecosystems. There is the clearest evidence that this is the case. However, if we examine the problem from a wider range of viewpoints we find that this is an issue of great complexity, even if the ecological problem is real and urgent.

What are the facts about deforestation? Between 0.5 and 1 per cent of the total of rainforest is being lost annually. This does not sound much, though it is more than 3 million hectares a year. Put another way, since the rate of forest clearance started to increase in the mid-1960s nearly a quarter has been lost. The initial clearance came with economic development of Amazonia following the completion of the road from Brasilia to Belem. Brasilia, the relocated capital of Brazil, is about 600 km from the densely populated coastal area, and Belem, 1,000 km north, is at the mouth of the Amazon. The road gave access to some of the parts of the world least affected by the modern world. Not for the first time better communications brought problems as well as benefits to humans. The benefits were economic development, the national priority in a relatively poor developing country. The forest land was converted to farm land, particularly for cattle ranching. The costs were the permanent loss of forest, the extinction of species and damage to the environment. Among the consequences of impact on the fragile tropical forest ecosystems were soil erosion and atmospheric impacts. In the case of the latter, this includes direct impacts such as the production of huge amounts of smoke from the clearance fires. Since the 1980s, in parts of Amazonia, commercial air travel has been disrupted for weeks at a time, due to poor visibility caused by smoke. The indirect effects have been a contribution to the buildup of CO_2 and a consequent worsening of the global greenhouse effect. Second, the loss of tree cover has altered the pattern of evapotranspiration in the Amazon basin. This has affected hydrology and regional climate, which in turn has acted upon vegetation in a feedback loop, and thus reinforced human impacts on the whole ecosystem. It is likely that a significant amount of the change to the environment and ecology of the area is permanent.

The government of Brazil promoted further development in Amazonia from the mid-1960s by financial incentives to investors, and by further improvements in infrastructure. Much of the development involved forest clearance for ranching. It is becoming clear that without continuing subsidy some of the farming schemes cannot continue. In some cases cleared land has been abandoned, while forest continues to be cleared. Thus some of the development is not viable *economically* as well as unsustainable *ecologically*. Future development of this area has to be based on a strategy for sustainability, in which biological conservation has a high priority. Protection of the rainforest ecosystems has the international highest significance.

However, before people in developed countries leap in with criticism, it is as well to review our own record. It is true that some of the developments in Amazonia are inappropriate by any criterion, and that there was little consideration to ecological priorities in the past. Some development was simply of the 'get-rich-quick', exploitive kind. This is changing albeit slowly. There have been problems, too, with inhabitants' human rights. But before we in the developed world condemn (and most of the readers of this book will see the world from a secure developed world perspective), consider our record in deforestation and land developments. Frankly, when reviewed over the past 500 years, it is not very good. Brazil was, and still is for many of its inhabitants, a relatively poor developing country. Economic growth remains the national priority. If the developed world wishes things to be different then perhaps we need to put more of our money where our mouths have been. This international problem requires international solutions. However, the ultimate solution to the problem, as well as the fundamental responsibility for the creation of the problem, remains with Brazil.

Large-scale impacts on ecosystems: some final thoughts

The very scale and complexity of large-scale human impacts on ecosystems makes their solution extremely difficult, the more so as in some cases human survival, driven by increasing population pressure, makes modification of resource management systems difficult. Yet for the main part we have a sufficient understanding of the ecology of impacts to know what to do. It is more generally a matter of translation of this knowledge into real and practical policies, at the core of which are sustainable development and international action on the ideas discussed at Rio in 1992. This meeting, if nothing else, brought the issues of human impacts on the planet into the core of international political activity. There is no time to lose because in some cases it is effectively impossible to reverse or restore damage completely. Global action is needed, and this will involve rich countries. This will involve all of us building a world for sustainable numbers of humans living sustainable lifestyles. At present the human ecological footprint is too heavy, and like an overused footpath is eroding its base. We must be sensitive to individuals' rights and freedoms, and to the realistic human desire for a good quality of life, but we must also realise that we all have a responsibility to sustain the ecological systems of the planet on which our existence depends.

Summary

- This chapter examines four types of large-scale human impacts on ecosystems.
- These are soil erosion, eutrophication and nitrate pollution, desertification and deforestation.
- All of these problems are found in many parts of the world, and no part of the populated world shows any symptoms of at least one of these issues.

Discussion questions

1   A significant cause of loss of forest cover in Africa is use of wood for fuel. Schemes to create sustainable forest projects that will provide continuing fuel supplies are being developed. Critically assess this type of forest project, considering issues such as alternative use of land and forest resources, environmental and economic costs of transport of wood fuel to cities and use of alternative fuels for cooking.

2   Recreational activities are an important economic use of water bodies. However, this use may cause ecological impacts that are harmful to water quality for other purposes. Should recreational activities be restricted to a limited number of water spaces? Should certain types of recreational use of water be banned in certain water bodies?

3   In what ways are the consequences of organic farming systems for soil resources an improvement on those resulting from intensive arable farming in developed countries?

Further Reading

See also

The role of disturbance in ecosystems, Chapter 6
Human impacts on ecosystems – impacts on trophic structure, Chapter 9
Global climatic change and ecosystems, Chapter 11

Further reading in Routledge Introduction to Environment Series

Natural Environmental Change
An Introduction to Sustainable Development
Environmental Policy
Environment and Society

General further reading

Changing the Face of the Earth (2nd edn). I.G. Simmons. 1996. Blackwell, Oxford.

This is still one of the best reviews of the whole subject, written eloquently yet showing dramatically the complexity of the history of effects of human societies upon the Earth.

Ecological Principles and Environmental Issues. P.J. Jarvis. 2000. Prentice Hall, Chichester.

This is an interesting and distinctive book that takes ecological theory and concept and applies these to a wide range of contemporary environmental problems. Chapter 3 (Ecosystem stability and chemical pollution) is most relevant to this chapter, but the whole book relates to the material in Chapters 9 to 11 of this book.

The Earth Transformed. A. Goudie and H. Viles. 1997. Blackwell, Oxford.

This is a clearly written and accessible review of human impacts. Although the whole book is relevant, Part 1 (Introduction to the Developing Environmental Impact) and Part 2 (The Biosphere) are especially useful.

11 Global environmental change: ecosystem response and biosphere impacts

The biosphere is in a constant state of change. The causes of change are strongly related to the functioning of ecosystems, in response to both internal and external factors. At a global level climatic change is the most important factor producing ecosystem change. Current concerns relate to human-induced global climatic change and the effects that this will have on the biosphere, and thus upon humankind. This chapter covers:

● **Global environmental change and its effect on the biosphere**
● **The key dimensions of global impacts on the biosphere**
● **Global climatic change and its effect on ecosystems**
● **Consequences of global climatic change for humans**

Assessing environmental change

This chapter is about change in the physical environment which controls ecosystems. Humans find change both exciting and threatening. For ecosystems, response to change is constant. The evolution of the planet, its ecosystems and life support systems is a record of continual change in which there are winners and losers. Biological winners are those organisms that evolve so successfully as to dominate ecosystems for long periods of time. Winners include flowering plants, particularly trees and grasses, and in the past, dinosaurs. Losers become extinct. So, as dinosaurs have shown, winners can become losers. This book has shown that the biotic and abiotic environment changes over time, and at different time scales, as well as in space at any point in time. In this section we examine global-scale environmental changes, which have had, and will continue to have, the ultimate influence on the biosphere as a whole. The principal control of environmental change as a whole is climate. In this chapter we consider how the biosphere may be changed as a result of global climatic change. This may be the most significant and threatening impact upon the biosphere which has ever confronted humankind.

Research shows that substantial change in the Earth's climate has been continuous over geological time. This is superimposed on short-term changes. This has been the environmental framework which has controlled ecosystem evolution and function since life first appeared on the planet. What is different about the human time period, a tiny fraction of the time during which life has been present, is that people have changed the environment more and more rapidly than at any time in the past, and humans have modified climate, as well as all other parts of the abiotic environment. It is not yet clear to what extent, and what will happen in the future, but more rather than less future environmental change, and thus impact on ecosystems, is certain, at least over the twenty-first century.

Global environmental change is normal. The dynamic nature of all biotic and physical environmental systems ensures that change is constant. Change in the environment is also very complex. However, the key **forcing factor** for all environmental change within the biosphere is climate. The exact processes of environmental change are not fully understood, so that laws are few, and predictions uncertain. There are two major reasons for this. First, environmental change involves many parameters. Even with the resources of modern science we are some considerable distance from complete analysis of all but the smallest of ecosystems. Furthermore, a serious problem for scientific analysis is that attempts to measure change in ecosystems, and the explanation of how these systems function, depend upon a complete knowledge of the environmental and biotic conditions at the start of the period under investigation. This poses one or both of two critical problems in studies of systems change. First, studies which are based on good-quality data are invariably short term, since accurate scientific observations are not available for more than about a hundred years at best, and generally for much lesser time periods. A considerable amount of scientific work has been devoted to the reconstruction of past environments and biological assemblages to rectify this gap in our knowledge and identify so-called **reference conditions** against which the degree of change may be assessed. Techniques such as analysis of sedimentary records, use of isotope dating methods and investigation of micro-fossil records have been developed with considerable skill and ingenuity. However, these records are incomplete and in many cases our knowledge of past environments and life is imprecise and incomplete.

The second reason why analysis of environmental change is very difficult is that it is the result of human actions, as well as natural agencies. As discussed in Chapter 8, a distinction between natural and human agencies is necessary because often the rate of change due to human impacts is more rapid than natural actions. Further, some types of human impacts occur naturally either not at all, or to little appreciable extent. It may often be difficult to distinguish between changes in the environment and in ecosystem functioning that are the result of natural processes, and those that are attributable to human impacts. Rates of natural change vary constantly, as the action of forcing factors varies over time. Our knowledge of the evolution of current ecological and environmental conditions is still limited. This might be taken as a reason to postpone the investigation of environmental change until scientific description of past environments is at a more advanced stage. However, problems related to human-induced environmental changes, which many believe are a significant threat to humankind and the biosphere, are too urgent to wait for knowledge to be painstakingly accumulated. We cannot wait for these problems to be broken down slowly, like a medieval siege. The urgent need is to gain an understanding of how we arrived at the current condition, and the role of human impacts in shaping the nature of change in environmental and ecological systems.

General characteristics of global impact on ecosystems

We can identify general characteristics that explain the ways in which humans have affected ecosystem behaviour, and the consequences of these impacts. In some cases, forcing factors that have a central role in environmental change, and thus effect upon ecosystem functioning can be identified. These are key factors. These can be considered similar to the keystone species, a designation used in conservation biology. Keystone species are considered to exert a powerful influence over the way in which an ecosystem functions, and thus protection or control of this species is critical in the overall conservation management of that ecosystem. The flying fox (*Pteropus sp.*) is a good example of a keystone species, or more properly genus, as there are many species of this animal (Primack 2002). Flying foxes are vital for plant pollination and seed dispersal throughout the islands of the Indian and Pacific Oceans. For some plant species, flying foxes are the only agents of pollination and seed dispersal. Therefore, decline and extinction of species of flying fox – a very real possibility for some species – would have a quite literally catastrophic impact on the entire ecosystems of these oceanic islands.

The action of key factors in environmental change results in changes to the normal pattern of system behaviour in both the biotic and abiotic elements of ecosystems. Changes in the energy budget of ecosystems are often important key factors. This is often related to climatic change. Variation in the input of solar radiation into ecosystems is a key forcing factor in environmental change. Thus decreased radiation input due to atmospheric pollution will reduce light and temperature, while buildup of **greenhouse gases** will cause increase in temperature. Change in atmospheric temperature is often associated with changes in atmospheric moisture. Heat and moisture conditions are among the most important factors influencing primary production in terrestrial ecosystems.

Changes in material cycles and budgets are a second general type of key factor in environmental change. Humans modify nutrient cycles and budgets through pollution, the causes of which may be deliberate (dumping wastes) or accidental (spills and leakage). Cycling is also affected by agricultural cropping of ecosystems with the consequent relocation of scarce nutrients. Impacts that affect the amount and type of organic debris, or the action of biological decomposers, will cause ecosystem damage. Intensive agriculture or sylviculture may also cause impacts on nutrient cycling.

Ecosystems are also affected by change to their biological components. A primary theme of this book is the dynamic character of functioning of the biological components of ecosystems: their plant, animal and microbial communities. Numbers of individual species vary through time, often quite short periods of time. In many environments, seasonal change of climate, whether it is alternation of warm and cold or wet and dry conditions, is a powerful influence on life. All organisms are the products of evolutionary processes. Since the dawn of life on the planet, species have evolved and become extinct, to be replaced in turn by newly evolved species. Evolution and extinction are natural and normal dimensions of ecosystem change. These processes take place over much longer time periods than the life of any individual. However, the recent increase in the range and amount of human impact on ecosystems has resulted in a correspondingly significant increase in rates of extinction among all types of biota. This has been caused in part by direct elimination of species by human actions, and in part by indirect change to the habitat of organisms. Finally, we must remember that impact on the biological component of ecosystems is an agent of change to the physical environment. There is reciprocal interaction between the biotic and abiotic parts of ecosystems. In Chapter 10 we examined problems associated with deforestation in Amazonia. Among these problems is the effect of deforestation on climate. Destruction of huge areas of tropical rainforest may influence the physical composition of the Earth's atmosphere through impact on the global carbon cycle, causing a buildup of CO_2. In this chapter, global climatic change is examined as a major cause of impacts on ecosystems throughout the biosphere. The issue of global climatic change shows that human modification of the environment may damage the reciprocal interaction between life on Earth and its physical environment, through the functioning of ecosystems.

The key dimensions of global impacts on the biosphere: people and climate

At the global scale there are two factors that dominate human impacts on ecosystems. There is a relationship between the two. The first factor is people, and more particularly the growing numbers of people and their greater effects on the biosphere through increased use of resources and new technologies. It could be argued that the ideas of T. Robert Malthus, who in 1803 in the extended version of his famous *Essay on The Principle of Population* contended that the size of the human population is controlled by the available food supply: the former growing geometrically while the latter could increase only arithmetically (Malthus 1803). The economic history of the world over the past two centuries has demonstrated that this view was not correct, at least over that time scale. However, the rate of resource depletion by the six billion humans alive today may yet show that the basic premise is sound. The second key factor is climate change. It is generally accepted scientifically that climate change as a result of human impacts on the atmosphere is taking place. The amount and rate of this change is uncertain but again it is believed that the amount of change will be significant, and its rate rapid. Change in the climate of Earth is important, since climate is a fundamental driver of environmental and ecological processes. Ecosystem functioning will be modified by climate change. Global climate change may thus be regarded as the critical outcome of human pressures, and a consequence of the growth in numbers of humans and their industrialised technologies, which have developed over the past 200 years. These two factors are now considered in more detail.

The human factor

Population numbers

There are a number of reasons for the acceleration in global-scale human impacts. It is important to consider these, because the development of strategies to prevent future damage, and amelioration of current effects on ecosystems depends on understanding human impacts. Analysis of the rate of impacts shows that most of the damage has occurred in the past 200 years and that the rate of change has accelerated during this time period. These are the centuries in which that monumental change in human activities known as the Industrial Revolution has taken place. The Industrial Revolution resulted

Table 11.1 World population growth since 1650

Year	Approximate world population	Average annual increase in numbers (from previous date)
1650	500 million	——
1820	1,000 million	2.94 million/year
1940	2,213 million	10.1 million/year
1960	3,020 million	40.3 million/year
1970	3,700 million	68 million/year
1980	4,500 million	80 million/year
1990	5,300 million	80 million/year
2000	6,200 million*	90 million/year
2010	7,100 million*	90 million/year
2025	8,300 million*	80 million/year

* = estimate

in huge human population growth. There are about six times as many people on the planet as there were 200 years ago. Table 11.1 shows the growth of world population since 1650. Population growth explosions of this order of magnitude in other animal species are normally followed by equally dramatic decline in numbers as **density-dependent** control factors operate. The degree of control that human beings can exert over their physical and biotic environment has so far prevented this. An example of the environmental consequences of industrialisation may be seen in increased energy use. This has been mainly through consumption of fossil fuels. Increased numbers of people and increased energy use have been accompanied by an enormous growth in the use of both biogeochemically renewable (flow) and non-renewable (stock) resources. The growth in use of natural resources has in turn led to problems associated with disruption of the functioning of natural cycle systems discussed in Chapters 2 and 3. Few parts of the biosphere have escaped substantial modification by humans, and over large tracts of the Earth's surface, especially in the developed world, ecological and environmental systems are, to a large extent, human artefacts.

Technology and resources

The increasing use of resources is a major part of the reason why the biosphere has been so altered by humans. To explain how and why these changes have taken place, it is important to understand what resources mean in a human context. A simple definition of resources is 'anything that is of use to man' (Porteous 1992). Box 11.1 examines the ways in which we can classify resources, and how this helps to understand the interaction between humans and the biosphere through resource utilisation. An important issue relating to technology is that both the amounts and the ways resources are used change over time. This makes understanding future demands on resources more difficult. A critical aspect for any resource which is renewable, such as resources derived from ecosystem function, is that whatever the level of use, it must be within the capability of the ecosystem to supply it with detriment to ecosystem functioning. In other words it must be sustainable.

In general terms human use of resources depends upon three conditions. First, resources will be required by humans to undertake some action which is deemed essential, useful or desirable by humans. Second, humans must have the technology to exploit resources to achieve the desired outcome. Technology may be defined as the knowledge required in order to apply resources to some purpose of human use, and more generally a function of accumulated human knowledge and socio-economic structures and goals. Third, resources have to be known to humans. Knowledge is the key to resource use. However, knowledge does not simply mean objective scientific knowledge, though this is a crucial element in technology. The goals of societies – economic growth, power over land and other people, or some religious or other value system – also affect how the resource base of the world is used. Some scientists are uncomfortable with ideas such as beliefs and value systems. This is because such notions are difficult to reconcile with the scientific approach to problems. However, as these humanistic issues have a considerable bearing upon environmental issues, it is impossible to ignore them. If we do forget them, we cannot understand the real nature of human impacts on the biosphere and still less can develop strategies to modify impacts so that the integrity of ecosystem function, which provides much of this resource base, can be developed. Box 11.2 looks at some of the broad issues relating to environment and society, and identifies some of the most important of these.

Box 11.1

Definition and classification of resources

Even the definition of resources is not as simple as it first seems. Implicit in the concept of resources is the idea that they are defined by humans. There is nothing in any material occurring on earth that makes it a resource without it being of some actual or potential use by humans. Linked to the idea of resources are population – the numbers of people – and technology – the ability of people to use resources.

Resources change in value over time. As they become scarcer, or as there are more uses to which they can be put, their value rises. As far as ecosystems are concerned the resources have three vital attributes:

- as food can be provided only by ecosystems, their resources are vital to all life, including human life
- ecosystem resources may be renewable, as a function of the continuous open energy-driven properties of the ecosystem
- ecosystem resources can be lost or disrupted by human impacts upon ecosystems and their functioning.

Resources can be classified in a number of ways, and these classification systems tell us a lot about the nature of resources. The most common classification is renewable and non-renewable resources. The former are derived from ecosystems, and their supply is maintained by ecosystem function. The latter are mineral and other resources which are derived from the abiotic environment, and are not renewable except at very long (i.e. geological) time scales. Non-renewable resources may be recycled by human process. These types of resources are also called stock and flow resources.

Within the classification we can further define stock resources which are consumed by use, theoretically recoverable or recyclable. Examples of these are fossil fuels, all elements in mineral form and metallic minerals respectively. Flow resources can be subdivided into critical zone resources (e.g. fish stocks) and non-critical zone resources (e.g. air) respectively. Flow resources depend upon ecosystem function, and the quality as well as the quantity of resource available is affected by human impact on ecosystems.

For a fuller discussion of the nature of resources see Rees (1990).

The Earth's atmosphere and climate change

The greatest human impacts on the global environment have been upon the atmosphere. As it is in the gaseous state, it is more dynamic than other parts of the physical environment. Its pivotal role in the hydrological, carbon and some macro-nutrient cycles means that changes to the atmosphere affect other environmental systems, and the biosphere. To a considerable extent climate is *the* forcing factor for ecosystems at a global scale, through supply of heat, water and its effect upon nutrient cycling. Therefore we shall examine some of the main issues in anthropogenic forced global climatic change through a review of three of the most significant components of atmospheric change caused by human

Box 11.2

Human impacts on the biosphere and societal values: a question of communications

It is an uncomfortable truth for scientists (or at least some scientists!) that eventually what seem like scientific problems become embroiled in human value systems. Human value systems may be irrational, and are generally hard to measure and quantify, but they form a central part of how most people live – the personal beliefs, religious or otherwise, the value people place on their material well-being, heritage and culture and so on. In the context of the problems of global environmental change, this has some important implications.

We have established that global climatic change will have significant consequences for the biosphere as a whole, and thus on the human resources which are provided by the biosphere. The whole of humankind is thus affected by these impacts. The sources of the impacts, however, are unequally contributed by different groups of people, both now and in the past. So though, at least in general terms, we understand the problems scientifically, and can use this knowledge to develop solutions, we have to be able to persuade all countries to subscribe to programmes of action. The Rio Earth Summit of 1992 and similar jamborees show just how difficult this is.

One of the key problems is that the poorer countries of the developing world may not wish to take action which they see as preventing them from attaining the material benefits already the property of the rich developed world. Some may be quite literally unable to afford environmental protection, others may allocate it a low priority in national development plans, while others may simply reject what they view as an attempt by the developed nations to retain their economic hegemony. It is hard to argue 'do not do what we did . . . and got rich in so doing'. On the other hand, developed countries are unwilling to contribute more than what they feel is their share to the solution of biosphere problems. Finding common ground is hard. And in a world increasingly dominated by democratic political systems, world leaders have to persuade their electorates that actions, which may hurt individuals economically in the short term, are good for all in the long term.

Fortunately, there is growing knowledge of and concern for the biosphere. This returns the international problem to the local arena. It will be here that real solutions to the human impact on the biosphere will be found, if at all. This is a compelling reason for more ecological and environmental research, the communication of research to all people and the heightening of environmental awareness as a central part of citizenship, in all countries in the world, whatever the level of development. We have argued in this book that the ecosystem concept is especially valuable in satisfying these aims. But just as much as the general public needs to know about the biosphere, environmental and ecological scientists must try to communicate with the non-scientific population, and be aware of people's beliefs, concerns and fears.

action. The relative significance of natural and human agencies of impact on ecosystems and the biosphere is also considered. This is a brief overview of the main issues.

A good starting point for literature on climate change is Drake (2000) and the websites http://www.pewclimate.org and http://europa.eu.int/comm/environment/climat/home, as well as the reading recommended at the end of this chapter.

Carbon dioxide in the atmosphere

Scientific data relating to atmospheric conditions during the recent past, as well as contemporary information, are relatively good. This is somewhat surprising, since the atmosphere is the most dynamic of all the spheres of the planet. However, its close functional links with other environmental systems and the biosphere mean that there is good evidence of past climatic conditions from a range of sources, ranging from the sedimentary record to micro-fossils. We also have reliable records of weather conditions – the state of the atmosphere at a particular point in time and space – at numerous locations over the surface of the Earth for periods of more than a hundred years. Weather recording stations are generally located in more populous and developed parts of the world, so that our knowledge of global conditions is less than ideal. It is not easy to detect long-term variations in weather patterns and climate scientifically, because both regular and irregular variations, often of considerable magnitude, occur as part of the normal functioning of atmospheric systems. Thus climate in most parts of the world varies in temperature and rainfall conditions on a seasonal basis, as well as less regularly from year to year. One of the major research problems is the identification of trends of change from a pattern of variation in parameters which change constantly. An increase in annual temperature of 2 or 3°C over a hundred years would indicate a significant change in climate. It is clearly difficult to identify such change when daily temperatures may vary by three or four times that amount.

However, a number of sets of records confirm that atmospheric composition has changed over the past hundred years (Elsom 1987). Atmospheric composition is closely linked to temperature. A summary of this is shown in Figure 11.1. These records show that there have been significant rises in carbon dioxide and methane in the atmosphere since 1800. The sources of these gases are related to human actions. Carbon dioxide has been produced in large quantities as a result of the combustion of fossil fuels. Methane is produced by a number of actions including decomposition of rubbish, natural decomposition processes in wetlands, cattle farming and combustion. Research has revealed that both carbon dioxide and methane concentrations were higher than at present in previous interglacial periods, indicating that such conditions can arise naturally. However, it is generally accepted that the rate of change over the past two centuries has been much faster than would occur as a result of natural systems behaviour alone (Mannion 1991). Carbon dioxide, methane and some other atmospheric components, which have also increased rapidly as a consequence of industrialisation, are greenhouse gases. These capture infra-red radiation (particularly the longer wavelength element of infra-red electromagnetic radiation) more efficiently than the shorter wavelengths. Radiation

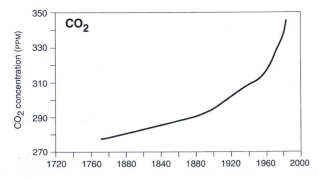

Figure 11.1 Changes in atmospheric carbon dioxide 1800 to 1980

re-emitted from the Earth is more to the longer end of the spectrum than that transmitted to the Earth by solar radiation. Thus these so-called greenhouse gases trap infra-red, that is heat radiation, as the troposphere acts like a greenhouse trapping radiation by means of its glass panes.

The way in which the greenhouse effect works is understood, and the evidence for buildup of greenhouse gases in the troposphere is clear. Although the scientific view is that climate change caused by humans has happened, the extent to which present climatic patterns have been modified cannot be specified precisely. Future changes in the Earth's climate are even harder to predict. Yet most scientists believe that the consequences for global climate as a result of the buildup of greenhouse gases are significant, and that there is a real risk that there will be resultant significant damage to the biosphere and to the planet's life-sustaining environmental and ecological systems.

Much of the greenhouse effect, at least at this time, is caused by an increase in tropospheric carbon dioxide concentration. The rapid consumption of fossil fuels, coal, oil and gas, upon which industrialisation depends, has acted to 'short-circuit' the delicate biologically maintained balances of the carbon cycle, described earlier. The buildup of carbon-based organic sediments, which took millions of years to accomplish, is being reversed over a few decades. There are other potential sources of CO_2 that might further boost atmospheric content. A very large pool of organic carbon is located in recent organic deposits, such as peat and mires. Should these stores break down, the consequences for climate would be considerable. The stability of stores is based on the condition and functioning of a number of rather fragile ecosystems.

The proportion of carbon dioxide and other greenhouse gases in the troposphere is small. Thus the increase in these gases represents a very small absolute change in the composition of the Earth's atmosphere: less than 0.1 per cent of the total of all gases involved. Yet the climatic effects of such a change will be considerable. Estimates of the heating that may occur over the next fifty years vary, but most scientists now see some increase in global temperature as inevitable. An increase of a degree or two centigrade would appear to be neither here nor there, but this is not so. Furthermore, some projections estimate an increase in global average temperature of as much as 6°C. Whatever the actual value of temperature increase is, the environmental and ecological consequences will be profound. First, there would be significant regional variations in the effects of global warming. Some areas might actually become cooler, especially in the winter, while other areas, particularly in the inter-tropical zone and continental interiors, may well have significantly higher temperatures (Schneider 1994). Second, it is not just temperature that is affected by global warming. A better term for the whole process is global climatic change. Patterns of rainfall and snow cover would be changed, and in some parts of the world there is likely to be an increase in the occurrence of extreme events such as storms and droughts. The delicate balance between atmosphere and hydrosphere that is responsible for the present oceanic current circulation pattern may be altered. The potential effects of such a change may be seen in the so-called El Niño event, in which ocean current patterns in the Pacific alter periodically, with considerable climatic consequences for much of the South American continent (Mannion 1991). It is possible that the warm Gulf Stream current, which at present grossly modifies winter temperatures in Northwest Europe, may be greatly reduced. An alternative effect may be that the increased level of CO_2 in the troposphere may boost levels of photosynthesis. It has even been suggested that the higher general metabolic rates which will result from increased temperature may be good for speciation and biodiversity. This is not likely however, as the rate of increase in temperature is rapid, and the negative effects of climate change will more than outweigh such benefits. The overwhelming body of scientific evidence supports the view that climate change will be a serious threat to biodiversity and ecosystem functioning. It is fair to note that there are some scholars who do not accept this. The recommended reading at the end of this chapter gives alternative perspectives on global warming.

Ozone depletion

The focus of attention on human-induced atmospheric change has been on the troposphere. However, the upper atmosphere has significance through its role in the radiation window of the Earth. Solar radiation potentially harmful to life is prevented from reaching the biosphere by the upper atmosphere. The ozone layer, located between 15 and 45 km above the Earth's surface, absorbs nearly all the ultraviolet radiation incident on the planet. Ultraviolet radiation is very harmful to plants, animals and microbiological organisms. Ozone (O_3) is formed by the photochemical disassociation of molecular oxygen (O_2) into atomic oxygen, which then combines with an oxygen molecule. Ozone not only blocks ultraviolet radiation transmission, but is also highly reactive chemically.

Since the early 1980s depletion of the ozone layer has been observed. Seasonal 'holes' in the ozone layer have appeared, first over the south Antarctic region, then in the same latitudes in the northern hemisphere. Damage to the ozone layer will allow higher levels of ultraviolet radiation to reach the Earth's surface. What is the cause of ozone depletion? It has been established that the chief culprit is a group of gases known as chlorofluorocarbons (CFCs). These manufactured gases are used as coolants in refrigeration systems, and in the past were widely used as propellants for **aerosol** sprays. The latter use has been prohibited in most developed countries. The former will have effect for some time to come. CFCs when released tend to migrate to the ozone layer, where they combine chemically with ozone. This has gone on at a rate faster than replacement by the photochemical synthesis of ozone. Depletion has resulted.

CFCs also have an effect as a greenhouse gas, as more radiation reaches the lower atmosphere. Thus in all ways the release of CFCs has been bad for the biosphere. It has been called most appositely a 'chemical weed' (E.P. Odum 1993). But even though we know that it is a problem, it is not easy to solve. Rich developed nations are more able than poor developing countries to abandon the use of CFCs. However, it is difficult to persuade, and impossible to coerce, developing countries into a course of action in which they will lose more than the already wealthy developed world. This issue raises the question of values in ecological and environmental resource use, which is discussed more fully in Box 9.3.

Increase in dust and aerosols

Dust and **aerosols** are the solid component of the atmosphere. The idea of a solid component of the atmosphere may seem a contradiction in terms, but in fact the troposphere contains a great deal of suspended solid particulate matter. All air contains dust and other particulate matter, including salt particles and fragments of organic debris and pollen. Human actions contribute considerably to the production of dust, and to generation of other solid materials, which enter the atmosphere (Elsom 1987). Dust particles and aerosols are so tiny that they may remain suspended in the atmosphere for long periods of time. If the air is in motion, not only can more material be transported as a result of the kinetic energy of the wind, but also the suspended load may be transported for considerable distances – hundreds of kilometres or more from their source areas. The occurrence of solid material in the atmosphere is, yet again, a natural process. In some cases very large amounts of dust may be suspended in the atmosphere. Dust storms in arid areas are well-known climatic hazards. In severe cases such storms can reduce visibility to less than 10 m, and an average storm will have a dust density of about 1 gm^{-3}. A further climatic impact of dust and aerosols is that it reduces incident radiation to the Earth's surface through scattering and absorbing heat and light.

Dust is a normal part of the environment. Fine soil particles, organic litter and salt crystals precipitated from sea water can be carried into the air by aeolian action. The large deposits of loess which are found bordering arid areas in East Asia and the Middle East are evidence of the scale of such movements, at times in the past million years or so. This happened because climatic conditions were favourable to aeolian transport, and there were abundant sources of material for transport. During the past 200 years human actions have accelerated the processes that load the atmosphere with dust. There are two ways in which this has happened. First, agricultural activity has exposed bare soil surfaces as a result of deflation by wind. This has been a serious problem in situations in which cultivation or grazing activities are carried out at levels beyond the carrying capacity of the resource base. The creation of the aptly named Dust Bowl of North America is the best known, but by no means the only example of the recent past. Second, humans have injected huge amounts of fine particulate matter into the atmosphere as a result of all forms of combustion; the smoke or exhaust produced as a result of burning all types of fossil fuel contain residues of solid material from the combustion. The amount and type of residue varies according to the type of fuel, the efficiency of the combustion process and any technological means which are employed to reduce emissions. The causes and effects of atmospheric pollution by particulate matter are discussed more fully in Box 11.3. The recent increase in aerosols is closely related to desertification, discussed in Chapter 10.

The effects of global climatic change on ecosystems

There are other impacts caused by humans, such as soil erosion or water pollution, which act on most parts of the planet, but climatic change has the most fundamental effect on the functioning of ecosystems. Therefore some of the impacts on the biosphere, which may result from global climatic change, are considered here. This evaluation will also consider some of the possible further consequences for the human life support base resulting from these impacts.

Global climatic change will have an effect on the patterns of temperature over the Earth's surface. Although there will be a general increase, there will be considerable regional variations, with some regions (such as the interior of the North American continent) experiencing larger increases than others (such as northern and western UK). Overall, the changes in temperature patterns are likely to produce spatial shifts in the location of ecosystems and even whole biomes (see Chapter 1). Vegetation communities, and the higher trophic levels in ecosystems, will tend to migrate pole-wards and/or upwards, as temperature increases. Scottish vegetation provides an example of what may occur. It should be remembered that the prediction for temperature increase in Scotland is lower than that in many other parts of the world, so that the consequent effect on ecosystems will be less. Increasing temperatures would effectively reduce the arctic-alpine climatic zones ("islands in the sky") located in the north and east of the country (Usher and Balharry 1996). This area has a more extreme and continental type of climate than any other part of Scotland, so that there is no location to which these ecosystems can migrate. Most Scottish ecosystems have been profoundly influenced by human actions over many centuries, but the least modified tend to be those at higher altitude.

Box 11.3

Atmospheric particulates and their effects on people and ecosystems

There have been considerable variations in patterns in time and space in smoke emissions throughout the world. The emissions were not a problem, except on a very local scale, until the industrial revolution. From the mid-eighteenth century, problems occurred in urbanised and industrialised areas of Europe and North America. The great smog – a deadly amalgam of meteorological fog and smoke from domestic and industrial sources – of London in 1952 is reckoned to have been directly responsible for the deaths of more than a thousand people; it was a catalyst to the enactment of the Clean Air Acts in the UK in the 1960s. By 1970 smoke emissions in London had been reduced to one-tenth of the level in 1956 (Goudie 1984: 304). This had a dramatic effect on emissions of particulate matter in cities, principally by restrictions on use of coal as a domestic fuel. Not only has this been beneficial to human health, but also cleaner city air has had benefits for plant life in urban areas. The problem has now moved to the developing world. For example severe smog problems are now being experienced in the rapidly growing cities of industrial China (Geping and Jinchang 1994: 146–7).

The effects of smog and other types of air pollution are harmful to all biota. Plants are sensitive to atmospheric pollution. **Stomata** become blocked by solid particles, and the chemical effects of atmospheric pollution, especially of sulphur dioxide, which is also a by-product of combustion, inhibit, damage or kill many species. Lichens are especially sensitive to such pollution, and have been used as biological indicators of the spatial extent and severity of atmospheric pollution (Elsom 1987). Some insects (such as the peppered moth, *Biston betularia*) developed industrial melanic forms to camouflage themselves on the black surfaces of trees and buildings. There are further climatic effects of particulate matter, and smog in polluted areas, which influence ecosystems. The reduction in radiation reaching the earth's surface should reduce temperatures, and indeed in the short term this is the case. However, as smog is generally a characteristic of urbanised areas, lower temperatures may be offset by the output of waste heat from space heating and other energy use in the city. This creates an urban **heat island**. In large cities in the developed world, winter heat output from urban activities can be 30 per cent or more of the solar radiation received. The higher concentration of dust caused by human activities, in and around cities, also affects rainfall. The tiny solid particles in the air form condensation nuclei around which water condenses to form rain drops. If more nuclei are available, and if sufficient atmospheric moisture condenses, this may increase rainfall. However, any potentially beneficially influences on climatic conditions in terms of more heat and moisture to support plant growth in cities are more than counteracted by the harmful effects of atmospheric pollution. Dust concentrations have an adverse effect on human health, and organic particles in particular are recognised as being harmful. Dust does not simply cause hay fever, but organic debris in the air is associated with more serious illnesses such as asthma and cancer. Thus humans have a more selfish reason to worry about the effects of atmospheric dust and aerosols, than an overall concern for impacts on ecosystem function. What this analysis of particulate matter in the atmosphere reveals is that the patterns of direct and indirect impacts of human-induced change to the atmosphere are complex.

Furthermore, some of these areas have biota and communities of European conservation significance. Therefore reduction in the extent of such areas, as a result of climatic change, has serious conservation implications. Even when communities survive in a modified climatic regime, some species may disappear (Dickinson 1995a).

Changes in moisture regime are likely to have even greater impacts on vegetation. Such impacts will be greatest in semi-arid locations, in which even a small absolute decrease in rainfall represents a large relative change. This will inevitably have major impacts on natural ecosystems as well as on human agricultural systems. It is not only the total amount of rainfall which will be affected by global climatic change, but also the degree of variability in annual rainfall. The potential agricultural consequences of changes in moisture regime, which may be produced as a result of global climatic change in Australia, are discussed by Russell (1988). A major issue in climatic change is the rapidity with which it seems to be taking place. Although the pattern is by no means clear, current research indicates that, while the amount of change in temperature and rainfall resulting from human-induced climatic change is similar to that which has occurred in the recent geological past, the rate of change is far more rapid than that resulting from natural system dynamics. Goudie (2000) provides a review of possible future climatic outcomes. Whatever the exact outcome, the result will be that many ecosystems will have great difficulty in migrating to an equivalent ecological niche quickly enough to cope with the changing climatic environment. Undoubtedly some species and possibly some whole communities will fail to relocate, and will become extinct. Finally we have the problem that, as yet, we have poor predictions of the way in which climate will vary at a local scale. There is evidence that such variations may be rapid and considerable. For example, annual rainfall in the west of Scotland has increased by as much as 30 per cent since the early 1970s (Dickinson 1995b). As the increase is concentrated in the winter months, the environmental as well as the ecological consequences of this change are likely to be substantial, with a higher incidence of storms and floods.

Consequences of environmental change

If the concept of the ecosystem is to be really useful for biological resource management, it must be able to provide a usable predictive model of the dynamics of ecosystem functioning. Although the extent to which this has been achieved may be questioned, the concept has considerable utility in its present state. An interesting case study, which evaluates the value of the ecosystem approach, and sets the approach in the context of conservation management policies, is provided by Wilcove (1994). He examined protection of the northern spotted owl (*Strix occidentalis caurina*) and old-growth forests in the Pacific Northwest region of the USA. He showed that to be successful, ecosystem-based conservation must take note of the whole range of species in the ecosystem, and their functional role in that ecosystem. One of the main uses of the ecosystem concept is the way in which the dynamic functioning of both the biotic and abiotic systems within the ecosystem can be related to causal factors. In this chapter we have seen how important external forcing factors may be in causing ecosystem change. But we have also seen how difficult it is to separate the components of natural change from those due to human impacts. It is these latter which cause humans so much concern; there is clear evidence that we have and continue to cause damage to ecosystems and their ability to support our existence. Therefore the ecosystem concept is of value in unravelling the relative contributions of natural and human agencies in ecosystem change, with the objective of understanding the former, and, where appropriate, managing the latter. The ecosystem structure is helpful, too, in identifying direct and indirect impacts of human actions. The latter in particular are difficult to predict. In some cases after the impact has occurred, indirect human impacts, such as changes to freshwater ecosystems following modified land use in the terrestrial part of the catchment area, remain hard to measure.

The practical value of the ecosystem concept is considerable; as already indicated, it is helpful in the identification of the consequences of human actions. Although it is not always possible as yet to develop satisfactory predictive models of ecosystems, progress has been made towards that end since 1980. Kitching (1983) showed how ecological modelling might be used as a predictive tool in developed countries such as the densely peopled nations of Europe. A particular use of ecosystems is in conservation management, which involves the manipulation of ecosystems that have been subject to millennia of human impact. Such ecosystems may contain biota, which are rare or endangered, or national heritage landscapes, so that the conservation value of such areas is high. As the case of the northern spotted owl shows, preservation alone is not a viable option. Understanding ecosystem function must form the basis of conservation management, and indeed all types of biological resource management throughout the biosphere.

Resource management for human purposes has been the main agent of impact on the biosphere. Biological resource management, such as agriculture or forestry, involves direct and deliberate manipulation of ecosystems. However, much of the impact that humans have had on ecosystems has not been deliberate. Examples, which have been examined in this book, include pollution and desertification. The scale of accidental human impacts on ecosystems varies from local to global. This chapter shows that the ecosystem approach provides the best way of understanding the complex results of human impacts. The ecological consequences of many human impacts remain poorly understood. There is a further problem. Even when a damaging impact is identified, people may choose to accept the damage to ecosystems. This will happen when people see impacts as being of lesser importance than economic gains resulting from resource use. It is important to understand that this is not an irrational response to the issue. In some countries, especially in the developing world, ecological and environmental impacts may be accepted as the inevitable consequence of economic growth. Rich developed countries may counsel that protection of ecosystems should have a higher priority than some developments. However, the past record during industrialisation in the rich nations was that environmental concerns had a low priority. To poor people it can seem that the rich wish to prevent the former from gaining what the latter already take for granted. This type of concern has been central to the political debate and actions since the Brandt and Brundtland reports (Brandt Commission 1980; World Commission on Environment and Development (WCED) 1987) and the Rio Earth Summit of 1992. There is no easy answer to this problem, but all people have the strongest incentive to resolve the issue – the survival of humankind. It is becoming clear that sustainability must be the key dimension of all human activities. This is essential to protect ecosystem functioning upon which the resources for human existence depend.

Conclusion: the value of the ecosystem concept in understanding the impacts of global environmental change

The ecosystem concept has been around since 1935. It has been criticised, but remains a central theme in ecology and environmental science. This book has shown that the ecosystem concept provides a useful framework for understanding the complex interactions that go on within the biosphere. In particular its value as a means of integrating the complex interactions between life and its environment is vital for the environmental sciences. The inclusion of human impacts in ecosystems analysis is a further strength.

The use of ecosystems does not preclude other paradigms, but rather complements other approaches. Particularly at large scales, the ecosystem provides the best way of understanding interaction and change. Chapters 8 and 9 have analysed how ecosystems are modified by human actions, and how ecosystems are responding to global-scale climatically forced change. The importance of these issues to humankind cannot be overemphasised. The recent spate of international conferences and pronouncements, sometimes accompanied by action from national governments, is a statement of the growing realisation that humans must be better stewards of the natural systems of the planet. Our survival depends upon this. But action must be based upon proper understanding of the problems, in turn based upon scientific knowledge. Understanding the world's ecosystems and their functioning is one step towards better understanding and better stewardship.

Global-scale problems require global-scale solutions. This does not mean that there is no place for more locally based scientific analysis, and for action by individuals and communities. Indeed it is likely that most advances will come at these scales. But some problems are global. The key problem, as this chapter has shown, is global climatic change and its effects on the biosphere. Its complexity is a continuing problem for science. The effects of change in climate on the complexity of the biosphere and its functioning are equally hard to predict. However, as the consequences for both biosphere and human societies are very likely to be substantial, we need to be able to develop better understanding of the likely outcome, and to develop global solutions to the problem. This means that science must be able to inform policy-makers. The debates that this issue produces are complex; yet science cannot shrink from this challenge. Further insights into these problems are given by O'Riordan (2000), Mitchell (2002) and O'Riordan (2004). The ecosystem concept provides a structure for the scientific analysis of organism–environment interactions and change. It also provides a structure whereby complex ideas can be communicated to non-scientists. For both of these reasons, the ecosystem concept, and its application to ecological and environmental problems, has a continuing importance for environmental and ecological scientists.

Summary

- This chapter examines the problem of global environmental change and its effects upon ecosystems.
- The human population of the planet has trebled in about a hundred years. This increase is placing great demands on the resource base, particularly of renewable resources that depend upon ecosystem functioning.
- The crucial global impact is human-induced climate change. The rates of change in climatic conditions that appear to be taking place are unprecedented in recent Earth history; the effects on the biosphere will be profound.
- The functioning and spatial location of many ecosystems will be changed. Biota, communities or even whole ecosystems may become extinct as a result of their inability to adapt to such rapid change.
- Some renewable human resources based upon ecosystems will become scarcer, and the effects of climatically forced change to ecosystems will affect all people.
- These global impacts require global-scale solutions which focus upon sustainable development.

Discussion Questions

1   Generating electricity by use of nuclear energy results in very low emissions of carbon dioxide. Should this type of electricity generation be encouraged? It is unlikely that renewable

forms of energy will be able to supply sufficient energy for current needs in most countries. Does this affect the potential use of nuclear power?

2 What are likely to be the main impacts of global climatic change on any local natural ecosystems with which you are familiar? Assess the impacts from the perspective of (a) increase in mean temperature, (b) change in rainfall, and (c) change in the frequency of occurrence of extreme climatic events, such as storms or droughts. Which biotas are likely to be most affected?

3 Will global climatic change have any effect upon the functioning of tropical ecosystems? If so, what changes, and why?

Further reading

See also

The role of disturbance in ecosystems, Chapter 6
Human impacts on ecosystems: impacts on trophic structure, Chapter 9
Large-scale impacts on ecosystems, Chapter 10

Further reading in Routledge Introduction to Environment Series

Biodiversity and Conservation
Natural Environmental Change
Environment and Society

General further reading

Global Environmental Issues: A Climatological Approach (2nd edn). David D. Kemp. 1994. Routledge, London.

This book is full of material relevant to this chapter, and remains a good source for climate change theory.

Global Warming: The Science of Climate Change. Frances Drake. 2000. Arnold, London.

A comprehensive but accessible review of the main issues involved in the scientific basis of global warming.

Our Changing Planet: An Introduction to Earth System Science and Global Environmental Change (3rd edn). Fred T. Mackenzie. 2000. Prentice Hall, Harlow.

This recent text links natural systems to human dimensions.

The Changing Global Environment. Neil Roberts (ed.). 1994. Blackwell, Oxford.

Authoritative series of essays on global environmental change by leading specialists in that field. Those by Roberts, Spencer, Dearing, Stott, Furley, Douglas and Goudie are particularly relevant to this chapter, but the whole book is of interest.

http://www.probiotech.fsnet.co.uk/savanna.htm

This website gives a summary of the views of one of the leading sceptics on the scale of global impacts caused by humans.

Glossary

abiotic   non-living, in the sense of the non-living part of an ecosystem.

adiabatic   with respect to atmospheric conditions this means without energy input from or output to an external source. Changes in the state of water between liquid and vapour mean that large amounts of energy are involved. Adiabic processes are central elements in weather systems.

aerosol   a mixture of very tiny particles of solid or liquid matter in the air. This is not a chemical combination, but because the particles are so tiny, aerosol particles can remain in the air for long periods. Smoke is an example of an aerosol, though clouds are not usually considered to be aerosols since the water droplets, which make up clouds, have become sufficiently large so as to move under the influence of gravity. Nearly all the aerosol particles in the atmosphere are located in the troposphere.

agro-ecosystems   ecosystems, generally simplified in structure, that are managed by humans to produce an output ('crop') for human use. The crop may be plant or animal material. Simplification is achieved by elimination of plant competitors ('weeds'), animal competitors (predators or 'vermin' and 'bugs') and plant parasites. This is increasingly carried through the action of agro-chemicals such as herbicides and pesticides. Agro-ecosystems are often based on disturbance-tolerant plants such as cereals, which are domesticated grasses. Agricultural tillage provides a suitable disturbed environment that favours their growth.

albedo   measure of the reflectivity of a surface expressed as the ratio of the radiation reflected by the surface to the total radiation incident on that surface.

allochthonous   in aquatic systems: 'external' primary production entering the system from land (e.g. as leaf litter from bankside trees).

anaerobic   deficient in oxygen (e.g. estuarine mud).

anthropogenic   describing ecological factors that are a result of human actions.

atmosphere   the shell of gases surrounding the surface of the Earth. The troposphere is that part closest to the Earth's surface, extending to about 10 km above the surface, and demarcated from layers above by a sharp change in the temperature gradient, containing about two-thirds of the total mass of atmospheric gases. The atmosphere is about 79 per cent N_2 and 20 per cent O_2. The small remaining part includes water vapour and CO_2, both of which are crucial to life and effectively confined to the troposphere. The Earth's atmosphere has evolved to its present conditions over geological time scales, partly as a consequence of interactions between life and its physical environment.

autochthonous   in aquatic systems: 'internal' primary production (e.g. from algae, cyanobacteria and aquatic macrophytes living in the water body).

autotrophic   photosynthetic or chemosynthetic organisms with the ability to use light or chemical energy to fix carbon into organic molecules usable as a food source.

available   when used to describe the condition of plant nutrients, this means in simple water-soluble ionic form, in the rooting zone. Thus they are in a condition which makes them available for use by plants. Much of the available pool of nutrients depends on cycling systems. Availability is affected by soil and climatic conditions.

biodiversity   this is not quite as straightforward to define as may at first seem. The Worldwide Fund for Nature (WWF) definition is as good as any: 'The millions of plants, animals and micro-organisms, the genes that they contain, and the intricate ecosystems they help build into the living environment.' In some ways the concept of biodiversity has no real scientific validity, being simply a shortening of the phrase 'biological diversity', but the ideas underpinning biodiversity have come to dominate contemporary biological conservation. Often associated with Edward Wilson, a giant of contemporary biological conservation, biodiversity is used in scientific writing to analyse the range of species and populations found in the biosphere, and interactions between these and the environment. Biodiversity became a global issue in the mid-1980s, and the convention on Biological Diversity, which came out of the 1992 Rio de Janeiro Earth Summit ensured that protection of biodiversity is a dimension in national and international governmental policies. Biodiversity has genetic, species and community dimensions and may be measured by α (alpha), β (beta) and γ (gamma) indices. The indices measure respectively the range of species (or other dimension) found in a specific location, the rate of change in species along an environmental gradient and the range of species found in a large area (subcontinental or continental scale), which will have a wide range of types of **habitats**.

biomass   the mass of an amount of biological material, generally applied to living organisms.

biome   a regional-scale assemblage of ecosystems, usually defined geographically (e.g. ocean biomes) or in terms of the dominant vegetation (e.g. rainforest biome).

biosphere   the narrow shell about the surface of the Earth, some 20 km thick and extending from the ocean abyss to the tropopause, within which all life is found.

biotic   living, associated directly with the living part of an ecosystem.

black box   a term used in system theory to describe a system, the internal functioning of which is unknown, but outputs and inputs to the system are known.

buffered   in a biological context this is the tendency to resist or to protect from change. For example, climate below the ground is less extreme, being buffered from the diurnal variations in temperature by overlying soil which transmits heat slowly. Buffer zones may be used as part of the protection strategy in some conservation areas. Buffering also describes a chemical process in which there is feedback that offsets change.

bulb   starch-rich, asexually produced regenerative organ in plants consisting of a short, usually vertical stem axis bearing a number of fleshy scale leaves.

CAM photosynthesis   a type of photosynthesis used by some desert plants that enables them to carry out photosynthesis without losing water through their stomata. CO_2 is absorbed at night and stored chemically (CAM = Crassulacean acid metabolism). This source of CO_2 is then used during daylight.

carrying capacity   the maximum size of a group of organisms which can be supported by a particular set of environmental conditions. Carrying capacity may be defined theoretically in the differential equation that is expressed graphically as the logistic or Verhulst-Pearl curve. The concept has been extended to some human uses of the environment, particularly outdoor recreation.

chelation   a natural process in the soil, in which cationic soil nutrients interact with organic acids. In strongly acid conditions iron may be removed from soil resulting in podsols (spodosols).

climax   in the context of the concept of Clementsian succession climax describes the final stable vegetation community, generally dominated by competitive species that result from the sequence of succession.

closed system   a system in which there are no movements of materials or energy across the defined system boundary. In ecosystems, most nutrient cycles may be regarded as being effectively closed systems.

community   an assemblage of populations of two or more species.

competition   effects of other organisms in competitive foraging for resources such as water, light, nutrients and space.

continuum   the ordered change in vegetation populations which occur along environmental and phytosociological gradients. Continuum analysis generally involves the family of statistical techniques known as ordination. An example of a widely used such technique in contemporary ecology is detrended correspondence analysis (DCA).

cryoturbation   substrate mobility caused by freeze–thaw action in soils, which regularly freeze then melt. Tundra soils often exhibit cryoturbation, and this produces highly disturbed conditions affecting plant root survival.

demostat model   a model in which populations are set at an equilibrium level as a result of environmental and population feedback controls.

density-dependent   this is a control factor which acts upon biological population growth in proportion to the density of that population. It is generally the most effective population regulation mechanism, since it is a negative feedback loop. Predation is an example of a density-dependent control.

desertification   the creation of desert-like conditions. Although this may be a result of natural climatic variations, more commonly over the past 100 years it has resulted from such human actions as overgrazing, loss of vegetation cover and over-cultivation.

disturbance   any environmental factor which damages or destroys the biomass of an organism, directly (e.g. for plants: grazing or forest fires) or indirectly by disturbing the organism's habitat (e.g. for plants: unstable substrate – such as a mountain scree slope).

dormancy   a phase in the life cycle of an organism in which development such as germination or reproduction is inhibited and overall biological metabolic rates are low. Dormancy occurs in many species of plants and animal species. Dormancy is an adaptation to cope with prolonged periods during which there are adverse environmental conditions, generally the climatic factors of limited availability of water or low temperatures.

ecological energetics   analysis of the amounts and flows of energy within ecosystems.

ecosphere   the biosphere together with the abiotic environmental systems which interact as ecosystems.

ecosystem 'an energy-driven complex of a community of organisms and its controlling environment' (Billings 1978).

entropy a measure of the heat energy of a body that is not available to do work. The concept helps us to understand that the fate of all energy is to end up as 'waste' heat, the least organised form of energy. Organisms use energy in many forms (e.g. chemical energy stores) but ultimately it all ends up as unusable heat.

eutrophication aptly described as too much of a good thing (Beeby and Brennan 2004), this process is a result of human modification of aquatic ecosystems. Primary biological production is limited in most freshwater systems by scarcity of macro-nutrients, particularly P and N. If soluble forms of these, particularly the former, are introduced into a freshwater body, then there will be an immediate response in a boost to primary production, generally by rapid growth of 'blooms' of algae. This change can be measured by an increase in the chlorophyll content of water. In extreme cases water will become deoxygenated as a consequence of the development of huge quant-ities of the algae, and all aerobic life will be extinguished. This risk is greatest in small, enclosed water bodies, in which atmospheric mixing is limited. The main sources of P and N, which are responsible for eutrophication, are poorly treated sewage effluent for P and synthetic fertilisers for N.

evapotranspiration the combined output of water vapour into the atmosphere from evaporation from the surface, and transpiration, the output of water vapour from plants, mainly associated with photosynthesis.

exergy a recent concept developed in **ecological energetics**: it is To(I), where To is the temperature of the environment and I is a measure of the 'thermodynamic informa-tion' of the system. It is effectively a measure of how far above the thermodynamic equilibrium (the state at which a system containing no living organisms would exist) the ecosystem is operating.

feedback in a system a path that carries information back to an earlier stage in the system modifying that **flow pathway**. Positive feedback amplifies flows and negative feedback reduces them. Negative feedback loops are thus often described as having a controlling function.

flow pathway in a system the path or flow of energy and/or materials. The functioning of flow pathways is responsible for system behaviour.

food chain the path of energy between different **trophic levels** in an ecosystem, through the actions of consumption and decomposition. At each step in the chain most energy is ultimately converted to heat through respiration and other chemical oxidations.

food web a more realistic description of the food chain. A web indicates the real world complexity in which most consumers consume different organisms, and that for the higher **trophic levels** consumption may be from different trophic levels.

forcing factor a causal agency or factor which is external to the functioning system under consideration. In the case of ecosystems, climatic variations are generally considered as forcing factors.

functional group an assemblage of populations of two or more species showing similar or analogous sets of traits for survival in the face of a defined set of pressures.

greenhouse gas any gaseous component of the troposphere which can absorb infra-red radiation from the sun or re-emitted from the Earth, more effectively than the main tropospheric gases nitrogen and oxygen. As these two gases make up nearly 99 per cent of the lower atmosphere greenhouse gases are a small atmospheric component. However, their environmental significance is considerable. Greenhouse gases, such as carbon dioxide and methane, occur naturally in the atmosphere, but their concentration has risen as a result of industrialisation. Greenhouse gases are a major element in human-induced global climatic change.

guild   a functional group of species sharing a common resource in **sympatry**, i.e. in such a way that their niches do not overlap.

habitat   an organism's 'address' (E.P. Odum 1953), i.e. the geographical location at which that organism lives, including the physical environmental characteristics of that location.

halophytic   in plants: tolerant of the stress associated with high salt concentrations in the environment.

heat island   area of positive temperature anomaly, normally in and around an urban area. The cause of the increase in temperature is heat energy added to the immediate environment from industrial processes and domestic heating.

heterotroph   organism requiring a supply of organic matter or food from the environment.

homiothermic   organisms with self-regulating temperature regimes. Commonly called warm-blooded, the burning of energy to produce heat is one of the most important homiothermic mechanisms. Such a metabolism requires very much larger energy intake by the organism than **poikilothermic** organisms.

hydrosphere   the shell of water that discontinuously covers about 70 per cent of the surface of the planet. This 'world ocean' makes up about 97 per cent of all the planet's water.

input   in the system context material or energy that goes into a system from beyond the system boundary.

intercalary meristems   growth tissue (meristematic tissue) in grasses, which is located along the stems of the plant. Such tissue allows growth from the base following removal of the upper parts of the plant by grazing or cutting. Most plants grow from apical meristems (at the tip or apex of a shoot or root) and thus regrowth following grazing is much slower, and generally must take place from a new shoot.

isotope   an alternative form of an element which is identical in chemical properties to the basic form, but which has a different composition of sub-atomic particles. For example, the 'normal' or common form of carbon is ^{12}C. A naturally occurring radioactive isotope is ^{14}C.

keystone species   in conservation biology, keystone species are identified because of the major influence such species have on overall ecosystem function. Keystone species may be found at any trophic level. Their action is generally via density-dependent controls, such as grazing or predation.

leaching   the removal of nutrients from the soil in solution in water draining through the soil.

lithosphere   the outermost shell of solid material making up the planet. The topmost part of this, the regolith, typically a few metres to tens of metres thick, is altered by the physical and chemical processes of weathering that break down and alter lithospheric material (rock). The very top of the **regolith** that is biologically active is the soil.

microbial loop   this refers to the ecosystem path in which dead organic matter is broken down by consumers such as protozoa. This is the key method of recycling nutrients.

model   a structured representation of reality. This may be in mathematical form, or a scale but real version of an actual system, an analogue model. An example of the latter is a wave tank that may be used to study the effects of ocean waves on coastal features. Mathematical models are widely used in ecology. They may also have predictive utility.

niche   a species 'profession' (E.P. Odum 1953) (see **habitat**). More properly, an abstract concept used to define that part of the ecosystem occupied by a given species: an *n*-dimensional volume with each of its n dimensions representing one ecological factor relevant to the survival of the species.

nutrients   the chemical elements that are essential for life. In other words, nutrients are the building blocks of life.

open system   a system in which there is movement of energy or materials across the system boundary. In ecosystems, the flux of solar energy to power life processes through photosynthesis, and the ultimate output of this energy through infra-red radiation from the earth, constitutes an open system.

operational functions   in a system those elements, paths and feedback loops that determine the operation of the system.

peds   soil aggregates ('clods, clumps' or more accurately prisms, columns, blocks and grains) that make up soil. Peds are formed by biological, chemical and physical processes within the soil, and are important elements influencing soil drainage and aeration.

permafrost   is literally permanently frozen ground. Occurring widely in the higher latitudes, especially of the northern hemisphere, which has large land area close above the arctic circle, permafrost has been formed by long-term contact with overlying ice. It is thus a relict of the last period of glacial advance. Permafrost only thaws out in the short polar summer to a very limited depth (a few tens of centimetres) and at deeper levels remains permanently frozen solid. Permafrost may be continuous, covering entirely huge tracts of land, or more sporadically located as discontinuous permafrost. Permafrost is very vulnerable to human impacts, and thus the stress-tolerant ecosystems that are able to cope with this challenging environment are also highly vulnerable.

phenological niche   that part of the ecological **niche** of a species which is defined by one or more factors related to timing of biological events (for example, plants which grow only early in the season in temperate woodlands, such as bluebells): the phenological factor(s) constitute one or more of the dimensions which define the niche of the species.

phytobenthos   derived from the Greek for 'plants of the bottom', but usually restricted to description of microscopic plants attached to substrate or other, larger, aquatic plants. Technically however all attached aquatic plants may be considered to be phytobenthos.

phytoplankton   derived from the Greek for 'plants which float', these are simple, microscopic, free-floating photosynthetic plants found in freshwater and marine aquatic environments.

phytosociology   the analysis of plant communities based on the premise that there is a 'sociology' of plants, i.e. that different species of plants grow together in more or less organised societies.

plant strategy theory   CSR theory. See **ruderals**.

poikilothermic   describes organisms that are dependent on their surroundings for the thermal environment in which their metabolism functions. The common term 'cold-blooded' does not make it clear that all plants are poikilothermic.

population   a breeding assemblage of individuals of a given species.

precautionary principle   if the risk to ecological conditions is so great that it cannot be accepted at any odds, then the action responsible for that risk is unacceptable, even if there is some degree of uncertainty about the actual outcome.

quadrat   a sample plot used for collecting information on vegetation. This information may record species present, the frequency of their occurrence in all quadrat samples, the space that each species occupies (cover) or the density of species' individuals in a quadrat. Quadrat means 'square' in German, but vegetation quadrats need not be square, though they commonly are. Size of quadrats varies according to purpose of survey, type of vegetation and data type collected. The commonest range of size is from 10×10 cm to 20×20 m.

rangelands   grass and scrub areas that are grazed by large animals, either wild (e.g. antelopes) or domesticated (e.g. cattle). Rangelands are semi-natural and their area worldwide has been extended by human use of fire and of the ranges for grazing. Both of these actions tend to favour disturbance-tolerant plant species over competitors. Grasses tolerate both fire and grazing well and thus dominate these areas. Rangelands are generally found in areas where there are significant water deficits. However, they are not simply a function of a climatic environment, and the role of humans and their ancestors over millennia is critical to their current extent and status.

reference conditions   the physical, chemical and biological conditions of an environmental system that apply at some specified system state. In ecological and environmental assessment this state may be defined by that in which there are no human impacts causing modification of the system. Reference conditions are used to establish the degree of human impact upon water bodies in monitoring regulations of the European Water Framework Directive (WFD) and thus define water quality.

ruderals   in plant strategy theory (CSR theory) plants, the ecological strategy which permits tolerance of disturbance. For a comparison of the characteristics of C (competitors), S (stress tolerant species) and R (ruderals) see Grime (2001: 89).

sere   stage in succession, identified by a distinctive plant community.

soil texture   refers to the proportion of different size classes (called soil fractions) of the mineral part of the soil. The different sizes of particles control the size of the spaces or voids in the soil. This is an important factor that influences soil water drainage and retention, and soil aeration.

specific heat   the amount of heat required to raise a given amount of matter through a specified temperature. Very simply we can think of it as a measure of the heat energy absorptive and storage capacity of a substance. It is generally measured in calories/gram (1 calorie = 4.187 joules). The temperature of a body measures the intensity of heat concentration; thus temperature will change by different amounts for materials that have different specific heat values. An environmentally important issue is that water has an extremely high specific heat (1 cal/gm), whereas earth surface materials are much lower (rock and soil are typically in the range 0.3–0.4 cal/gm). This means that water heats up and cools down much more slowly than land surfaces.

stochastic   describes a system with probabilistic elements in its functioning. A stochastic system does not have a simple determined trajectory, and it is impossible to predict its functioning in a way that does not describe different probabilities of outcomes.

stomata   pores on the leaves of photosynthetic plants which allow air from the atmosphere to enter the photosynthetic tissues, providing a source of CO_2. When stomata are open, loss of water occurs from plants through transpiration.

stratosphere   the atmospheric layer above the **troposphere**, which starts at about 10 km above the Earth's surface at the tropopause. Although largely devoid of life it effects the functioning of the troposphere by acting as a thermal lid on the troposphere. This is a result of the increase in temperature progressively above the tropopause.

stress   any environmental factor affecting an organism's physiological efficiency and hence survival ability; for example, in plants stress (e.g. shade) limits the ability to accumulate C through photosynthesis, thereby reducing productivity.

succession   the sequence of development of vegetation starting from a sterile surface. Each stage in the succession is known as a **sere**, which is characterised by a distinctive assemblage of plants. As succession proceeds towards the final stable end-point or **climax**, the communities tend to become less dominated by stress and disturbance-tolerant species, and dominated by species with highly competitive ecological strategies. The concept of the climax was first developed by Clements. E.P. Odum has attributed several other characteristics to succession. Some, such as increasing biological productivity, are widely accepted, whereas others, such as self-regulation, remain controversial.

support functions   the processes within an ecosystem that support the life present within it.

sustainable development   'development which meets the needs of the present without compromising the ability of future generations to meet their own needs' (World Commission on Environment and Development 1987).

sympagic   describes ecosystems closely associated with the undersides of floating sea ice in the Arctic and Antarctic Oceans. Dominated by invertebrates and microbiota these ecosystems are important elements in the prolific marine life of these oceans.

sympatry   refers to related and similar species living in the same geographical area and which diverge genetically by speciation. Sympatric speciation fulfils the requirement that species cannot occupy identical ecological **niches**.

trophic level   the feeding location of an organism. In other words, its location in the food chain. Photosynthetic plants are at the first or primary trophic level, grazers at the second level, and consumers of grazing animals at the tertiary and so on.

trophic structure   the structure of energy transfer and loss between populations in the ecosystem.

tropopause   the boundary between the troposphere and the atmospheric layer above, the **stratosphere**. The tropopause is marked by a change in thermal gradient. In the troposphere temperature decreases with distance from the Earth's surface, while in the stratosphere it increases with altitude.

troposphere   the atmospheric layer closest to the Earth's surface. It makes up about 70 per cent of the total mass of the atmosphere.

user   in an ecological sense an organism that makes use of any part of an ecosystem. Humans are users *par excellence*.

Bibliography

Abernethy, V.J., McCracken, D.I., Adam, A., Downie, L., Foster, G.N., Furness, R.W., Murphy, K.J., Ribera, I., Waterhouse, A. and Wilson, W.L. 1996. 'Functional analysis of plant–invertebrate–bird biodiversity on Scottish agricultural land', in *The Spatial Dynamics of Biodiversity*, I.A. Simpson and P. Dennis (eds). Proceedings of 5th Annual IALE Conference, Stirling 1996, 51–9.

Adams, C. 1994. 'The fish community of Loch Lomond; its history and rapidly changing status', in *The Ecology of Loch Lomond,* K.J. Murphy, M.C.M. Beveridge and R. Tippett (eds). Kluwer, London.

Agostinho, A.A., Thomaz, S.M., Minte-Vera, C.M. and Winemiller, K.O. 2000. 'Biodiversity in the high Paraná River floodplain', in *Biodiversity in Wetlands: Assessment, Function and Conservation*, B. Gopal, W.J. Junk and J.A. Davis (eds). Backhuys, Lieden.

Ali, M.M., Dickinson, G. and Murphy, K.J. 2000. 'Predictors of plant diversity in a hyperarid desert wadi ecosystem,' *Journal of Arid Environments* 45, 215–30.

Appleton, J. 1975. *The Experience of Landscape*. Wiley, Chichester.

Attenborough, D. 1979. *Life on Earth*. Collins and BBC, London.

Beeby, A. and Brennan, A-M. 2004. *First Ecology*. Oxford University Press, Oxford.

Billings, W.D. 1978. *Plants and the Ecosystem*, 3rd edn. Wadsworth, Belmont, CA.

Binns, T. 1990. 'Is desertification a myth?', *Geography* 75, 106–13.

Bonkowski, M. 2004. 'Protozoa and plant growth: the microbial loop in soil revisited', *New Phytologist* 162, 617–31.

Bowler, P.J. 1992. *The Fontana History of the Environmental Sciences*. Fontana, London.

Bradbury, I. 1991. *The Biosphere*. Belhaven, London.

Brandt Commission. 1980. *North–South: A Programme for Survival*. Pan, London.

Briggs, J., Dickinson, G., Murphy, K., Pulford, I., Belal, A.E., Moalla, S., Springuel, I., Ghabbour, S.I. and Mekki, A-M. 1993. 'Sustainable development and resource management in marginal environments: natural resources and their use in the Wadi Allaqi region of Egypt', *Applied Geography* 13, 259–84.

Burke M.J.W. and Grime, J.P. 1996. 'An experimental study of plant community invasibility'. *Ecology* 77, 776–790.

Busso, C.A., Bóo, R.M. and Pelaez, D.V. 1993. 'Fire effects on bud viability and growth of *Stipa tenuis* in semiarid Argentina', *Annals of Botany* 71, 377–81.

Campbell, D.G. 1992. *The Crystal Desert*. Martin, Secker & Warburg, London.

Caulfield, C. 1982. *Tropical Moist Forests*. Earthscan, London.

Chase, M.W. (and 41 others). 1993. 'Phylogenetics of seed plants: an analysis of nucleotide sequences from the plastid gene *rbc*L', *Annals of the Missouri Botanical Garden* 80, 528–80.

Clements, F.E. 1916. 'Plant succession: analysis of the development of vegetation', *Carnegie Institute Washington Publication* 242, 1–512.

—— 1928. *Plant Succession and Indicators*. Hafner Press, New York.

—— 1936. 'Nature and the structure of the climax', *Journal of Ecology* 24, 252–84.

Cloud, P. and Gibor, A. 1970. 'The oxygen cycle', *Scientific American* 223, 59–68.

Cloudsley-Thompson, P.L. 1989. 'Desertification or sustainable yields from arid environments', *Environmental Conservation* 15, 197–204.

Cohen, J.E., Briand, F. and Newman, C.M. 1990. *Community Food Webs: Data and Theory*. Springer-Verlag, New York.

Colinvaux, P. 1980. *Why Big Fierce Animals are Rare*. Penguin, Harmondsworth.

—— 1993. *Ecology 2*. Wiley, New York.

Connell, J.H. and Slayter, R.O. 1977. 'Mechanisms of succession in natural communities and their role in stabilization and organization', *American Naturalist* 1119–1144.

Countryside Commission for Scotland. 1992. *The Mountain Areas of Scotland*. Countryside Commission for Scotland, Battleby, Perth.

Cowles, H.C. 1899. 'The ecological relationships of the vegetation on the sand dunes of Lake Michigan', *Botanical Gazette* 27, 95–117; 167–202; 281–301; 361–91.

Crawford, R.M.M. 1989. *Studies in Plant Survival*. Blackwell, Oxford.

Cutter, S.L. and Renwick, W.H. 1999. *Exploitation, Conservation, Preservation: A Geographical Perspective on Natural Resource Use*, 3rd edn. Wiley, Chichester.

Deevey, E.S. 1970. 'Mineral cycles', *Scientific American* 223, 148–58.

Diaz, S., Hodgson, J.G., Thompson, K., Cabido, M., Cornilissen, J.H.C., Jalili, A., Montserrat-Marti, G., Grime, J.P., Zarrinkamar, F., Asri, Y., Band, S.R., Basconelo, S., Castro-Diez, P., Funes, G., Hamxchee, B., Khoshnevi, M., Pérez-Rontomé, Shirvany, F.A., Vendramini, F., Yazdani, S., Abbas-Azimi, R., Bogaard, A., Boustai, S., Charles, M., Dehghan, M., de Torres-Espuny, L., Falczuk, V., Guerrero-Campo, J., Hynd, A., Jones, G., Kowsary, E., Kazemi-Saeed, F., Maestro-Martinez, M., Romo-Diez, A., Shaw, S., Siavash, B., Villar-Salvador, P. and Zak, M.R. (2004) 'The plant traits that drive ecosystems: Evidence from three continents', *Journal of Vegetation Science* 15, 295–304.

Dickinson, G. 1988. 'Countryside recreation', in P. Selman (ed.) *Countryside Planning: The Scottish Experience*. Stirling University Press, Stirling.

—— 1995a. 'Conservation and environmental change', *Journal of the Association of Scottish Geography Teachers* 24, 38–42.

—— 1995b. 'Environmental impacts in the Loch Lomond area of Scotland', in H. Coccissis and P. Nijkamp (eds) *Sustainable Tourism*. Avebury, Aldershot.

—— 2000. 'Recreation at Scottish Lochs'. *Journal of the Association of Scottish Geography Teachers* 29, 41–51.

Dickinson, G., Murphy, K.J. and Springuel, I. 1994. 'The implications of the altered water regime for the ecology and sustainable development of Wadi Allaqi, Egypt', in A.C. Millington and K. Pye (eds) *Environmental Change in Drylands: Biogeographical and Geomorphological Perspectives*. Wiley, Chichester.

Dobson, M. and Frid, C. 1998. *Ecology of Aquatic Ecosystems*. Harlow, Longman.

Drake, F. (2000) *Global Warming: The Science of Climate Change*. London, Arnold.

Eleftherohorinos, I.G., Murphy, K.J. and Drennan, D.S.H. 1985. 'Weed competition from *Galium aparine* L. and *Veronica persica* Poir. in winter wheat and their control by chlorsulfuron', *Aspects of Applied Biology 9: The Biology and Control of Weeds in Cereals*, 197–203. Association of Applied Biologists, Warwick, UK.

Elsom, D. 1987. *Atmospheric Pollution: Causes, Effects and Control Policies*. Blackwell, Oxford.

Elton, C. 1927. *Animal Ecology*. Macmillan, London.

Embley, R.W., Baker, E.T., Chadwick, W.W., Lupton, J.E., Resing, J.A., Massoth, D.J. and Nakamura K. 2004. 'Exploration of Mariana Arc volcanoes reveal new hydrothermal systems', *EOS* 85, 37–44.

Farmer, A.M. and Spence, D.H.N. 1986. 'The growth strategies and distribution of isoetids in Scottish freshwater lochs', *Aquatic Botany* 26, 247–58.

Ferguson, N. 2003. *How Britain Made the Modern World*. Allen Lane, London.

Flanagan, D. (ed.). 1970. *The Biosphere*. Freeman, San Francisco, CA.

Fryer, J.D. and Chancellor, R.J. 1970. 'Herbicides and our changing weeds', in *The Flora of a Changing Britain*. BSBI Report 11, London.

Furley, P.A. and Newey, W.W. 1983. *The Geography of the Biosphere*. Butterworth, London.

Gaston, K.J. (ed.). 1996. *Biodiversity: A Biology of Numbers and Difference*. Blackwell, Oxford.

—— and Spicer. J.I. 1998. *Biodiversity: An Introduction*. Blackwell, Oxford.

Gaudet, C.L. and Keddy, P.A. 1988. 'Predicting competitive ability from plant traits: a comparative approach', *Nature* 334, 242–3.

Gause, G.F. 1932. 'Experimental studies on the struggle for existence. I. Mixed population of two species of yeast', *Journal of Experimental Biology* 9, 389–402.

—— 1934. *The Struggle for Existence*. Williams and Wilkins, Baltimore, MD.

Gemmell, R.P. 1982. 'The origin and botanical importance of industrial habitats', in R. Bornkamm, J.A. Lee and M.R.D. Seaward (eds) *Urban Ecology*. Blackwell, Oxford.

Geping, Q. and Jinchang, L. 1994. *Population and the Environment in China*. Paul Chapman, London.

Gimingham, C.H. 1972. *Ecology of Heathlands*. Chapman and Hall, London.

Gleason, H.A. 1926. 'The individualistic concept of the plant association', *Torrey Botanical Club Bulletin* 53, 7–26.

Godwin, H. 1977. 'Sir Arthur Tansley: the man and the subject', *Journal of Ecology* 65, 1–26.

Gordon, N.D., McMahon, T.A., Finlayson, G. and Nathan, R.J. (2004). *Stream Hydrology: An Introduction for Ecologists*, 2nd edn. Chichester, Wiley.

Goudie, A.S. 1984. *The Nature of the Environment*. Blackwell, Oxford.

—— (ed.). 1997. *The Human Impact Reader*. Blackwell, Oxford.

—— (2000). The *Human Impact*, 5th edn. Blackwell, Oxford.

Goudie, A. and Viles, H. 1997. *The Earth Transformed*. Blackwell, Oxford.

Gould, S.J. 1980. *Ever Since Darwin: Reflections on Natural History*. Pelican, London.

Grace, J. 2004. 'Understanding and managing the global carbon cycle', *Journal of Ecology* 92, 189–202.

Grainger, A. 1993. 'Rates of deforestation in the humid tropics: estimates and measurements', *Geographical Journal* 159, 33–44.

Gray, M. 2004. *Geodiversity: Valuing and Conserving Abiotic Nature*. Wiley, Chichester.

Grime, J.P. 1973. 'Competitive exclusion in herbaceous vegetation', *Nature* 242, 344–7.

—— 1979. *Plant Strategies and Vegetation Processes*. Wiley, New York.

—— 2001. *Plant Strategies, Vegetation Processes, and Ecosystem Properties*. Wiley, Chichester.

—— Hodgson, J.G. and Hunt, R. 1988. *Comparative Plant Ecology*. Unwin Hyman, London.

Hardin, G. 1960. 'The competitive exclusion principle', *Science* 131, 1292–7.

Hills, J.M. and Murphy, K.J. 1996. 'Evidence for consistent functional groups of wetland vegetation across a broad geographical range of Europe', *Wetlands Ecology and Management* 4, 51–63.

Hills, J.M., Murphy, K.J., Pulford, I.D. and Flowers, T.H. 1994. 'A method for classifying European riverine wetland ecosystems using functional vegetation groups', *Functional Ecology* 8, 242–52.

Hodgson, J.G. 1991. 'Management for the conservation of plants with particular reference to the British flora', in I.F. Spellerberg, F.B. Goldsmith and M.G. Morris (eds) *The Scientific Management of Temperate Communities for Conservation*. Blackwell, London.

Hooijer, A. 1996. *Floodplain Hydrology*. Proefschrift, Vrije Universiteit Amsterdam.

Houghton, J.T., Meira Filho, L.G., Callander, B.A., Harris, N., Kattenerg, A. and Maskell, K. (eds) (1996) *Climate Change 1995: The Science of Climate Change*. Cambridge University Press, Cambridge.

Hutchinson, G.E. 1957. *A Treatise on Limnology, Volume 1: Geography, Physics and Chemistry*. Wiley, New York.

Jarvis, P.J. 2000. *Ecological Principles and Environmental Issues*. Prentice Hall, Chichester.

Jeffers, J.N.R. 1978. *An Introduction to Systems Analysis with Ecological Applications*. Edward Arnold, London.

Jorgensen, S.E. 1992. *Integration of Ecosystem Theories: A Pattern*. Kluwer, Dordrecht.

—— 1996. 'The application of ecosystem theory in limnology', *Verhandlungen der Internationale Vereinigung für theoretische und angewandte Limnologie* 26, 181–92.

Keddy, P.A. 1989. *Competition*. Chapman and Hall, London.

Kemp, D.D. 1994. *Global Environmental Issues: A Climatological Approach*, 2nd edn. Routledge, London.

Kitching, R.L. 1983. *Systems Ecology: An Introduction to Ecological Modelling*. Queensland Press, London.

Koblentz-Mishke, O.J., Volkovinsky, V.V. and Kabanove, J.G. 1970. 'Plankton primary production of the world ocean', in W.S. Wooster (ed.) *Scientific Exploration of the South Pacific*. National Academy of Sciences, Washington, DC.

Lachavanne, J.B. 1985. 'The influence of accelerated eutrophication on the macrophytes of Swiss lakes: abundance and distribution', *Verhandlungen der Internationalen Vereinigung für theoretische und angewandte Limnologie* 22, 2950–5.

Lack, D. 1947. *Darwin's Finches*. Cambridge University Press, Cambridge.

—— 1954. *The Natural Regulation of Animal Numbers*. Oxford University Press, Oxford.

Lake, J.C. and Leishman, M.R. 2004. 'Invasion success of exotic plants in natural ecosystems: the role of disturbance, plant attributes and freedom from herbivores', *Biological Conservation* 117, 215–226.

Lévêque, C. and Monunoou, J-C. 2003. *Biodiversity*. Wiley, Chichester.

Li, Y. and Norland, M. 2001. 'The role of soil fertility in invasion of Brazilian pepper (*Schinus terebinthifolius*) in Everglades National Park, Florida', *Soil Science* 166, 400–5.

Liddle, M.J. 1975. 'A selective review of the ecological effects of human trampling on natural ecosystems', *Biological Conservation* 7, 17–36.

—— and Scorgie, H.R.A. (1980). 'The effects of recreation on freshwater plants and animals: a review', *Biological Conservation* 17, 183–206.

Liebig, J. 1840. *Chemistry and its Application to Agriculture and Physiology*. Taylor and Walton, London.

Lindeman, R. 1942. 'The trophic–dynamic aspect of ecology', *Ecology* 23, 399–418.

Lomborg, B. 2001. *The Skeptical Environmentalist*. Cambridge University Press, Cambridge.

Loreau, M., Naeem, S. and Inchausti, P. (eds). 2002. *Biodiversity and Ecosystem Functioning*. Oxford University Press, Oxford.

Lotka, A.J. 1925. *Elements of Physical Biology*. Williams and Wilkins, Baltimore, MD.

Loucks, O.L. 1962. 'Ordinating forest communities by means of environmental scalars and phytosociological indices', *Ecological Monographs* 37, 137–66.

Lovelock, J. 1988. *The Ages of Gaia*. Oxford University Press, Oxford.

MacArthur, R.H. 1958. 'Population ecology of some warblers of northeastern coniferous forests', *Ecology* 39, 599–619.

MacArthur, R.H. and Wilson, E.O. 1967. *The Theory of Island Biogeography*. Princeton University Press, Princeton, NJ.

Mackenzie, F.T. 2000. *Our Changing Planet: An Introduction to Earth System Science and Global Environmental Change*, 3rd edn. Prentice Hall, Harlow.

McKinney, M.L. and Schoch, R.M. 2003. *Environmental Science: Systems and Solutions*, 3rd edn. Jones and Bartlett, London.

McQueen, D.J., Johannes, M.R.S., Post, J.R., Stewart, T.J. and Lean, D.R.S. 1989. 'Bottom-up and top-down impacts on freshwater pelagic community structure', *Ecological Monographs* 59, 289–309.

Malthus, T.R. 1803. *An Essay on the Principle of Population: A View of its Past and Present Effects on Human Happiness; With an Inquiry into Our Prospects Respecting the Future Removal or Mitigation of the Evils which It Occasions*, 2nd revised edn. John Murray, London.

Mann, K.H. 1996. *Dynamics of Marine Ecosystems: Biological–Physical Interactions in the Ocean*, 2nd edn. Blackwell, Oxford.

Mannion, A.M. 1991. *Global Environmental Change: A Natural and Cultural Environmental History*. Longman, Harlow.

—— (1999) *Natural Environmental Change*. Routledge, London.

—— and Bowlby, S.R. (eds). 1992. *Environmental Issues in the 1990s*. Wiley, Chichester.

Margalef, R. 1968. *Perspectives in Ecological Theory*. University of Chicago Press, Chicago, IL.

Marsh, G.P. 1864, republished 1965. *Man and Nature; Or Physical Geography as Modified by Human Action*. Harvard University Press, Cambridge, MA.

Mitchell, B. 2002. *Resource and Environmental Management*, 2nd edn. Prentice Hall, Harlow.

Möbius, K. (1877). *Die Auster und die Austernwirtschaft*. Berlin.

Moss, B. 1988. *Ecology of Fresh Waters: Man and Medium*, 2nd edn. Blackwell, Oxford.

Murphy, K.J. 1983. 'Herbicide-resistance in weeds: a growing problem', *Biologist* 30, 211–19.

—— 2002. 'Plant communities and plant diversity in softwater lakes of Northern Europe', *Aquatic Botany* 73, 287–324.

—— Rørslett, B. and Springuel, I. 1990. 'Strategy analysis of submerged lake macrophyte communities: an international example', *Aquatic Botany* 36, 303–23.

—— Dickinson, G., Thomaz, S.M., Bini, L.M., Dick, K., Greaves, K., Kennedy, M., Livingstone, S., McFerran, H., Milne, J., Oldroyd, J. and Wingfield, R. 2003. 'Aquatic plant communities and predictors of diversity in a sub-tropical river floodplain the Upper Rio Paraná, Brazil', *Aquatic Botany* 77, 257–276.

Myers, N. (ed.). 1985. *The Gaia Atlas of Planet Management*. Pan, London.

Newson, M. 1992. *Managing the Human Impact on the Natural Environment*. Wiley, Chichester.

Nillson, C. and Grelsson, G. 1995. 'The fragility of ecosystems: a review', *Journal of Applied Ecology* 32, 677–692.

Odum, E.P. 1953. *Fundamentals of Ecology*. Saunders, Philadelphia, PA.

—— 1971. *Fundamentals of Ecology*, 3rd edn. Saunders, Philadelphia, PA.

—— 1983. *Basic Ecology*. Saunders, Philadelphia, PA.

—— 1993. *Ecology and our Endangered Life Support Systems*, 2nd edn. Sinauer, Sunderland, MA.

Odum, H.T. 1983. *Systems Ecology: An Introduction*. Wiley, Chichester.

O'Riordan, T. 2000. *Environmental Science for Environmental Management*, 2nd edn. Harlow, Prentice Hall.

—— 2004. 'Environmental science, sustainability and politics', *Transactions of the Institute of British Geography* NS 29, 234–47.

Padisak, J. 1993. 'The influence of different disturbance frequencies on the species richness, diversity and equitability of phytoplankton in shallow lakes', *Hydrobiologia* 249, 135–56.

Park, T. 1954. 'Experimental studies of interspecies competition. II. Temperature, humidity and competition in two species of *Tribolium*', *Physiological Zoology* 27, 177–238.

Phillips, J.D. 1992. 'The end of equilibrium?', in J.D. Phillips and W.H. Renwick (eds) *Geomorphic Systems: Proceedings of the 23rd Binghampton Symposium in Geomorphology*. Elsevier, Amsterdam.

Pickering, K.T. and Owen, L.A. 1994. *An Introduction to Global Environmental Issues*. Routledge, London.

Pieterse, A.H. and Murphy, K.J. 1993. *Aquatic Weeds*, 2nd paperback edn. Oxford University Press, Oxford.

Pimental, D. and Pimental, M. 1979. *Food, Energy and Society*. Wiley, New York.

Porteous, A. 1992. *Dictionary of Environmental Science and Technology*. Wiley, Chichester.

Primack, R.B. 2002. *Essentials of Conservation Biology*, 3rd edn. Sinauer, Sunderland, MA.

Pulford, I.D., Murphy, K.J., Dickinson, G., Briggs, J.A. and Springuel, I. 1992. 'Ecological resources for conservation and development in Wadi Allaqi, Egypt', *Botanical Journal of the Linnaean Society* 108, 131–41.

Pullin, A.S. 2002. *Conservation Biology*. Cambridge University Press, Cambridge.

Ranwell, D.S. 1972. *Ecology of Salt Marshes and Sand Dunes*. Chapman and Hall, London.

Rees, J.A. 1990. *Natural Resources: Allocation, Economics and Policy*, 2nd edn. Routledge, London.

Reynolds, C.S. 1996. 'The plant life of the pelagic', *Verhandlungen der Internationale Vereinigung für theoretische und angewandte Limnologie* 26, 97–113.

Ricklefs, R.E. 1990. *Ecology*, 3rd edn. Freeman, New York.

Roberts, N. (ed.). 1994. *The Changing Global Environment*. Blackwell, Oxford.

Robinson, R.A. and Sutherland, W.J. 2002. 'Post-war changes in arable farming and biodiversity in Great Britain', *Journal of Applied Ecology* 39, 465–77.

Rodwell, J.S. (ed.). 1991a. *British Plant Communities. Volume 1. Woodlands and Scrub*. Cambridge University Press, Cambridge.

—— (ed.). 1991b. *British Plant Communities. Volume 2. Mires and Heath*. Cambridge University Press, Cambridge.

—— (ed.). 1992. *British Plant Communities. Volume 3. Grassland and Montane Communities*. Cambridge University Press, Cambridge.

—— (ed.). 1993. *British Plant Communities. Volume 4. Aquatic Communities, Swamps and Tall-herb Fens*. Cambridge University Press, Cambridge.

—— (ed.). 2000. *British Plant Communities. Volume 5. Maritime Communities and Vegetation of Open Habitats*. Cambridge University Press, Cambridge.

Russell, J.S. 1988. 'The effect of climatic change on the productivity of Australian agroecosystems', in G.I. Pearman (ed.) *Greenhouse: Planning for Climatic Change*. CSIRO (Commonwealth Scientific and Industrial Research Organisation), Melbourne.

Sandquist, G.M. 1985. *Introduction to System Science*. Prentice Hall, Englewood Cliffs, NJ.

Schneider, S.H. 1994. 'Detecting climatic change signals: are there any fingerprints?', *Science* 263, 341–347.

Schroder, H.K., Andersen, H.E. and Kiehl, K. 2005. 'Rejecting the mean: estimating the response of fen plant species to environmental factors by non-linear quantile regression', *Journal of Vegetation Science* 16, 373–82.

Sidorkewicj, N.S., López Cazorla, A.C., Murphy, K.J., Sabbatini, M.R., Fernández, O.A. and Domaniewski, J.C.J. 1997. Interaction of common carp with aquatic weeds in Argentine drainage channels', *Journal of Aquatic Plant Management* 36, 5–10.

Simmons, I.G. 1996. *Changing the Face of the Earth*, 2nd edn. Blackwell, Oxford.

Tansley, A.G. 1935. 'The use and abuse of vegetational concepts', *Journal of Ecology* 16, 204–307.

—— 1949a. *Britain's Green Mantle*. Allen & Unwin, London.

—— 1949b. *The British Islands and their Vegetation*. Cambridge University Press, Cambridge.

Tarbuck, E.J. and Lutgens, F.K. 1997. *Earth Science*, 8th edn. Prentice-Hall International, London.

Thomas, D.S.G. and Middleton, N.J. (1994) *Desertification: Exploding the Myth*. Wiley, Chichester.

Thomas, W.L. (ed.). 1956. *Man's Role in Changing the Face of the Earth*. University of Chicago Press, Chicago, IL.

Tilman, D. and Kilham, S.S. 1976. 'Phosphate and silicate growth and uptake kinetics of the diatoms *Asterionella formosa* and *Cyclotella meninghiniana* in batch and semi-continuous culture', *Journal of Phycology* 12, 375–383.

Tivy, J. 1971. *Biogeography: A study of Plants in the Ecosphere*. Oliver & Boyd, Edinburgh.

Tolba, M.K. and El-Kholy, O.A. (eds). 1992. *The World Environment, 1972–92*. Chapman and Hall, London.

Turner, B.L., Clark, W.C., Kates, R.W., Richards, J.F., Mathews, J.T. and Meyer, W.B. (eds). 1990. *The Earth as Transformed by Human Action*. Cambridge University Press, Cambridge.

Usher, M.B. 1973. *Biological Management and Conservation*. Chapman and Hall, London.

—— and Balharry, D. 1996. *Biogeographical Zonation of Scotland*. Scottish Natural Heritage, Perth.

Walter, M.R. (ed.). 1996. *Evolution of Hydrothermal Ecosystems on Earth (and Mars?)*. CIBA Foundation Symposium 202. Wiley, Chichester.

Warren, A., and Agnew, C. 1988. 'An assessment of desertification and land degradation in arid and semi-arid areas', *International Institute for Environment and Development (IIED), Drylands Programme, Paper 2*, IIED, London.

Westoby, M., Falster, D.S., Moles, A.T., Vesk, P.A. and Wright, I.J. (2002). 'Plant ecological strategies: some leading dimensions of variation between species', *Annual Review of Ecology and Systematics*. 33, 125–60.

White, G. 1789, republished 1971. *The Natural History of Selborne*, Everyman edn. Dent, London.

Whitmore, T.C. 1998. *An Introduction to Tropical Rain Forests*, 2nd edn. Oxford University Press, Oxford.

Whittaker, R.H. 1956. 'Vegetation of the Great Smokey Mountains', *Ecological Monographs* 26, 1–80.

Wilcove, D.S. 1994. 'Turning conservation goals into tangible results: the case of the spotted owl and old-growth forests', in P.J. Edwards, R.M. May and N.R. Webb (eds) *Large-scale Ecology and Conservation Biology*. Blackwell, Oxford.

Williams G.H. 1982. *Dictionary of Weeds of Western Europe*. Elsevier, Amsterdam.

Wilson, E.O. 1994. *The Diversity of Life*. Penguin, Harmondsworth.

Wilson, S.D. and Keddy, P.A. 1986. 'Species competitive ability and position along a natural stress/disturbance gradient', *Ecology* 67, 1236–42.

Wilson, W.L., Abernethy, V.J., Murphy, K.J., Adam, A., McCracken, D.I., Downie, I.S., Foster, G.N., Furness, R.W., Waterhouse, A. and Ribera, I. 2003. 'Prediction of plant diversity response to land use change on Scottish agricultural land', *Agriculture Ecosystems and Environment* 94, 249–63.

World Commission on Environment and Development (WCED) (Brundtland Commission). 1987. *Our Common Future*. Oxford University Press, Oxford.

Websites cited

http://europa.eu.int/comm/environment/climat/home
http://www.pewclimate.org
http://www.sepa.org.uk
http://www.snh.org.uk

Index

Note: page numbers in *italic* denote references to figures/tables.

eBooks

eBooks – at www.eBookstore.tandf.co.uk

A library at your fingertips!

eBooks are electronic versions of printed books. You can store them on your PC/laptop or browse them online.

They have advantages for anyone needing rapid access to a wide variety of published, copyright information.

eBooks can help your research by enabling you to bookmark chapters, annotate text and use instant searches to find specific words or phrases. Several eBook files would fit on even a small laptop or PDA.

NEW: Save money by eSubscribing: cheap, online access to any eBook for as long as you need it.

Annual subscription packages

We now offer special low-cost bulk subscriptions to packages of eBooks in certain subject areas. These are available to libraries or to individuals.

For more information please contact webmaster.ebooks@tandf.co.uk

We're continually developing the eBook concept, so keep up to date by visiting the website.

www.eBookstore.tandf.co.uk